A2-Level

Biology

Edexcel

The Revision Guide

Editors:
Ellen Bowness and Simon Little.

Contributors:
Ben Aldiss, Gloria Barnett, Wendy Butler, Martin Chester, James Foster, Derek Harvey, Kate Houghton, Kate Redmond, Katherine Reed, Rachel Selway, Adrian Schmit, Emma Singleton, Jennifer Underwood.

Proofreader:
Sue Hocking.

Published by Coordination Group Publications Ltd.

ISBN-10: 1 84146 396 5
ISBN-13: 978 1 84146 396 4
Groovy website: www.cgpbooks.co.uk
Jolly bits of clipart from CorelDRAW®
Printed by Elanders Hindson Ltd, Newcastle upon Tyne.

Contents

Unit 4: Respiration and Coordination and Options

Section One — Respiration

Section Two — Physiological Control

Option A

Section Three — Microbiology and Biotechnology

Option B

Section Four — Food Science

Option C

Section Five — Human Health and Fitness

Unit 5: Genetics, Evolution and Biodiversity

Section Six — Photosynthesis

Section Seven — Biodiversity

Section Eight — Genetics and Evolution

Section Nine — Genetic Engineering

Energy and the Role of ATP

All animals and plants need energy for life processes and also for reading books like this. This stuff is pretty tricky and we're diving in at the deep end, so hang on...

Biological processes need **Energy**

Cells need **chemical energy** for biological processes to occur. Without this energy, these processes would stop and the animal or plant would just **die**... not good.

Energy is needed for **biological processes** like:
- active transport
- muscle contraction
- maintenance of body temperature
- reproduction and growth (i.e. cell division)

Plants need energy for **metabolic reactions**, like:
- photosynthesis
- taking in minerals through their roots

ATP carries **Energy** around

It only **gets worse** from here on in for the rest of the section. But **don't worry**, ATP didn't make any sense to me at first — it just clicked after many **painful hours** of reading dull books and listening to my teacher going on and on. Here goes...

1) ATP (**adenosine triphosphate**) is a **small water-soluble** molecule that is easily transported around cells.
2) It's made from the nucleotide base **adenine**, combined with a **ribose sugar** and **three phosphate groups**.
3) ATP is a **phosphorylated nucleotide** — this means it's a nucleotide with extra phosphate groups added.
4) ATP **carries energy** from **energy-releasing** reactions to **energy-consuming** reactions.

How ATP carries energy:

1) ATP is **synthesised** from **adenosine diphosphate** (**ADP**) and an **inorganic phosphate** group using the energy produced by the **breakdown of glucose**. The enzyme **ATPsynthase** catalyses this reaction.
2) ATP **moves** to the part of the cell that requires energy.
3) It is then **broken down** to ADP and **inorganic phosphate** and **releases chemical energy** for the process to use. **ATPase** catalyses this reaction.
4) The ADP and phosphate are **recycled** and the process starts again.

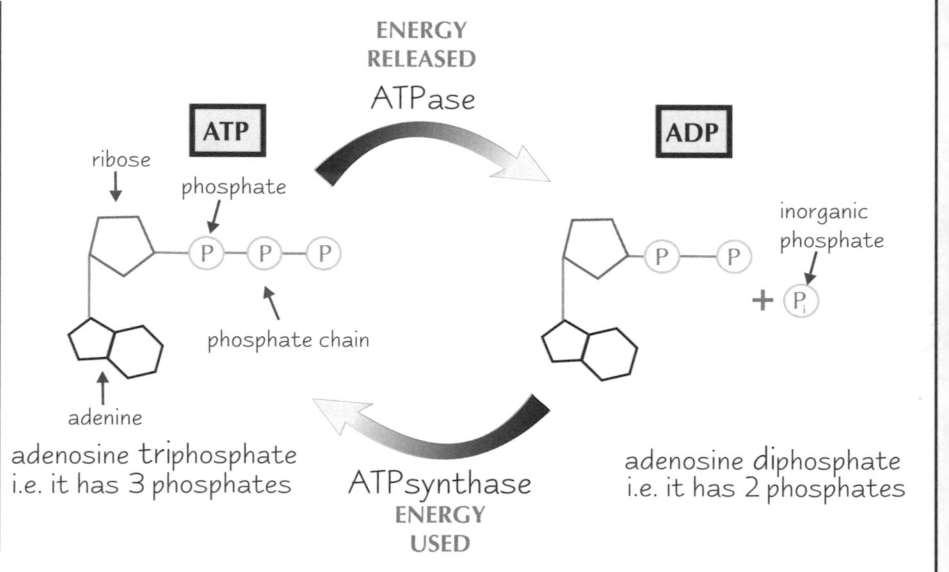

Cells Release **Energy** (to make **ATP**) by **Respiration**

Cellular respiration is the process where cells **break down glucose**. It releases **energy**. The energy is used to **produce ATP** from ADP and P_i. There are two types of respiration:

1) **Aerobic respiration** — respiration **using oxygen**.
2) **Anaerobic respiration** — respiration **without oxygen** (see page 8 for more). Both types produce ATP.

You need to learn the summary equation for **aerobic respiration**.

$$C_6H_{12}O_6 \text{ (glucose)} + 6O_2 \longrightarrow 6CO_2 + 6H_2O + \textbf{Energy}$$

Energy and the Role of ATP

Respiration takes place in the Mitochondria of the Cell

1) **Mitochondria** are present in all **eukaryotic** (i.e. plant, animal, fungi and protoctist) cells. They're 1.5 to 10 μm long.
2) Cells that use lots of energy, e.g. **muscle cells**, **liver cells** and the middle section of **sperm**, have lots of mitochondria.
3) The **inner membrane** of each mitochondrion is folded into **cristae** — structures that increase surface area.
4) **ATP** is produced via the **stalked particles** on the cristae of the inner mitochondrial membrane, in a stage called the **electron transport chain** (see page 7).
5) The **Krebs cycle** and the **link reaction** (pages 5-6) take place in the **matrix** of mitochondria.

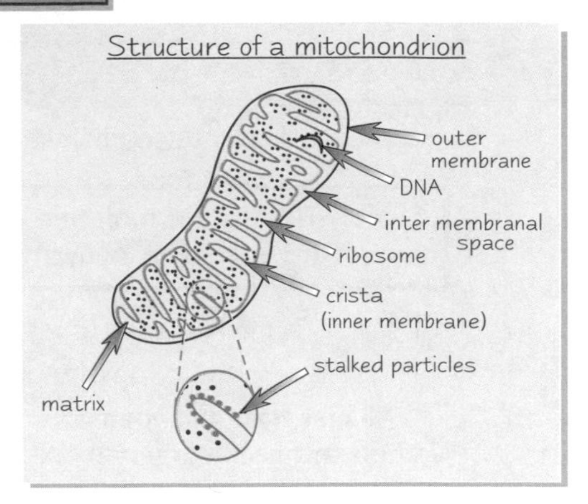

Respiration is a Metabolic Pathway

There are some pretty confusing technical terms about reactions in this section. If you do chemistry you'll be laughing — if not, you'd better concentrate:

- **Metabolic pathway** — a **series** of **small reactions**, e.g. respiration or photosynthesis, where each step is controlled by specific enzymes.
- **Catabolic reactions** — **breaking large molecules** into **smaller ones** using enzymes, e.g. breaking down glucose in respiration.
- **Anabolic reactions** — **combining smaller molecules** to make **bigger ones** using enzymes.
- **Phosphorylation** — **adding phosphate** to a molecule, e.g. ADP is phosphorylated to ATP.
- **Hydrolysis** — the **splitting** of a molecule using **water**. The enzymes that catalyse these reactions are called **hydrolases**.
- **Photolysis** — the **splitting** of a molecule using **light** energy.

Redox reactions — reactions that involve **oxidation** and **reduction**.

1) If something is **reduced** it has **gained electrons**, and lost oxygen or gained **hydrogen**.
   If something is **oxidised** it has **lost electrons**, and gained oxygen or lost **hydrogen**.
2) Oxidation of one thing always involves reduction of something else.
3) The enzymes that catalyse redox reactions are called **oxidoreductases**.
4) Respiration and photosynthesis are riddled with redox reactions.

Oxidation	Reduction
electrons are lost	electrons are gained
oxygen is added	oxygen is lost
hydrogen is lost	hydrogen is gained

One way to remember electron movement is "OILRIG" = Oxidation Is Loss of e^-, Reduction Is Gain of e^-.

Practice Questions

Q1 How is energy released from ATP?

Q2 Write down five metabolic processes in animals which require energy.

Q3 What is the purpose of the cristae in mitochondria?

Q4 What are oxidoreductases?

Exam Questions

Q1 What is the connection between phosphate and the energy needs of a cell? [2 marks]

Q2 ATP is a small, water-soluble molecule which can be rapidly and easily converted back into ADP if ATPsynthase is present. Explain how these features make ATP suitable for its function. [3 marks]

I've run out of energy after that little lot...

You really need to understand what ATP is, because once you start getting bogged down in the complicated details of respiration and photosynthesis, at least you'll understand why they're important and what they're producing. It does get more complicated on the next few pages, so take your time to understand the basics before you turn the page.

Glycolysis

You can split the process of respiration into four parts — that way you don't have to swallow too many facts at once. The first bit, glycolysis, is pretty straightforward.

Glycolysis is the *First Stage* of *Respiration*

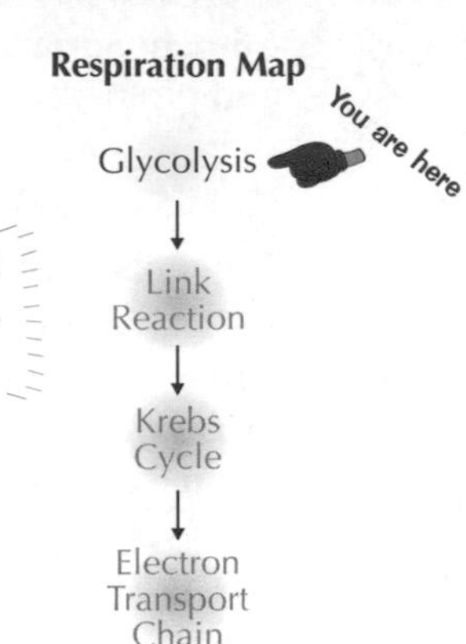

So, to recap... most cells use carbohydrates, usually glucose, for respiration.

> Glycolysis splits **one molecule** of glucose into **two** smaller molecules of **pyruvate**.

Glucose is a hexose (6-carbon) molecule.
Pyruvate is a triose (3-carbon) molecule.
Pyruvate is also known as pyruvic acid.

1) Glycolysis is the first stage of respiration (see the map to the right).
2) It takes place in the **cytoplasm** of cells.
3) It's the **first stage** of both aerobic and anaerobic respiration, and **doesn't need oxygen** to take place — so it's **anaerobic**.

There are *Two Stages* of *Glycolysis* — *Phosphorylation* and *Oxidation*

1 Stage One — Phosphorylation

1) Glucose is **phosphorylated** by adding 2 **phosphates** from 2 molecules of ATP.
2) Glucose is split using water (**hydrolysis**).
3) 2 molecules of **glycerate 3-phosphate (GP)** and 2 molecules of ADP are produced.

2 Stage Two — Oxidation

1) The glycerate 3-phosphate is **oxidised** (loses hydrogen), forming **two** molecules of **pyruvate**.
2) **Coenzyme NAD⁺** collects the hydrogen ions, forming **2 reduced NAD** (**NADH + H⁺**).
3) **4 ATP** are produced, but 2 were used up at the beginning, so there's a **net gain** of **2 ATP**.

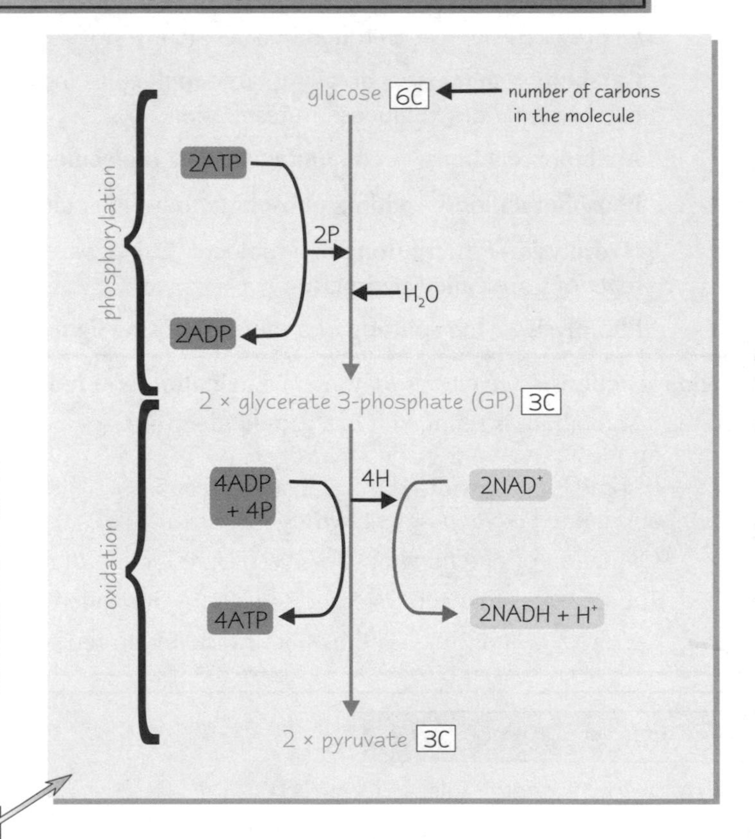

A coenzyme is a helper molecule that carries chemical groups or ions about, e.g. NAD removes H⁺ and carries it to other molecules.

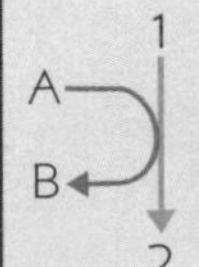

These arrows in diagrams just mean that A goes into the main reaction and is converted to B. A will normally release or collect something from molecule 1, e.g. hydrogen or phosphate.

Next in *Aerobic Respiration...*

1) The **2** molecules of **reduced NAD** go to the **electron transport chain** (see page 7).
2) The **two pyruvate** molecules go in to the matrix of the **mitochondria** for the **link reaction** (a small reaction that **links** glycolysis to the second stage, the **Krebs cycle**). It's so exciting I bet you can't wait...

Link Reaction

Glycolysis

The Link Reaction converts Pyruvate to Acetyl Coenzyme A

The link reaction is fairly simple and goes like this:

1) One **carbon atom** is removed from pyruvate in the form of $\mathbf{CO_2}$.
2) The remaining **2-carbon molecule** combines with **coenzyme A** to produce **acetyl coenzyme A** (**acetyl CoA**).
3) Another oxidation reaction occurs when $\mathbf{NAD^+}$ collects more **hydrogen ions**. This forms **reduced NAD** ($\mathbf{NADH + H^+}$).
4) **No ATP** is produced in this reaction.

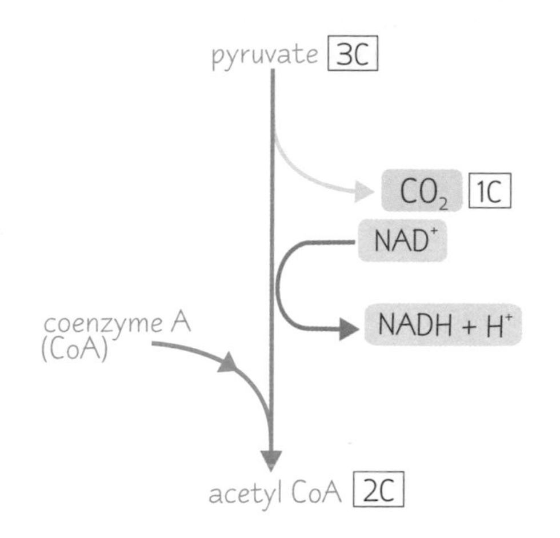

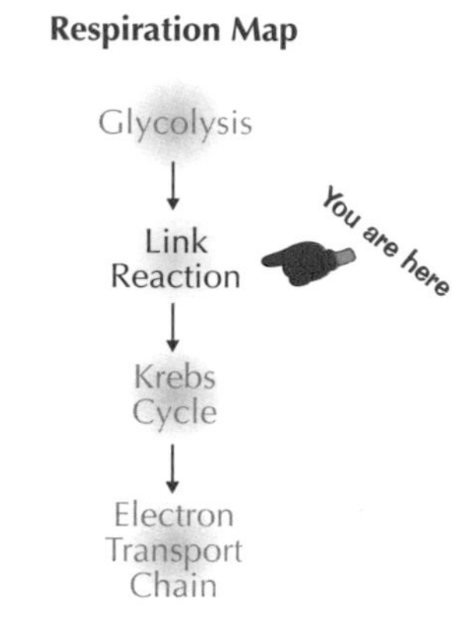

The Link Reaction occurs Twice for every Glucose Molecule

1) For each **glucose molecule** used in glycolysis, **two pyruvate** molecules are made.
2) But the **link reaction** uses only **one pyruvate** molecule, so the **link reaction** and the **Krebs cycle** happen **twice** for every glucose molecule which goes through glycolysis.

The Products of the Link Reaction go to the Krebs Cycle and the ETC

So for each glucose molecule:

- Two molecules of **acetyl coenzyme A** go into the Krebs cycle (see next page).
- Two **carbon dioxide molecules** are released as a waste product of respiration.
- Two molecules of **reduced NAD** are formed and go into the **electron transport chain** (which is covered on the next two pages).

Practice Questions

Q1 What do the terms hydrolysis and phosphorylation mean?

Q2 Why is there only a net gain of 2 ATP during glycolysis?

Q3 What is the final product of glycolysis?

Q4 Where is acetyl CoA formed?

Exam Questions

Q1 Describe simply how a 6-carbon molecule of glucose can be changed to pyruvate. [5 marks]

Q2 Describe what happens in the link reaction. [4 marks]

Acetyl Co-what?

It's all a bit confusing, but you need to know it, so it's worth taking a bit of time to break it down into really simple chunks. Don't worry too much if you can't remember all the little details straight away. If you can remember how it starts and what the products are, you're getting there. You'll get the hang of it all eventually, even if it seems hard right now.

Krebs Cycle and Electron Transport Chain

And now we have the third and fourth stages of the respiration pathway. Keep it up — you're nearly there.

*The **Krebs Cycle** is the **Third Stage** of Aerobic Respiration*

The Krebs cycle takes place in the **matrix** of the mitochondria. It happens once for each pyruvate molecule made in glycolysis, and it goes round twice for every glucose molecule that enters the respiration pathway.

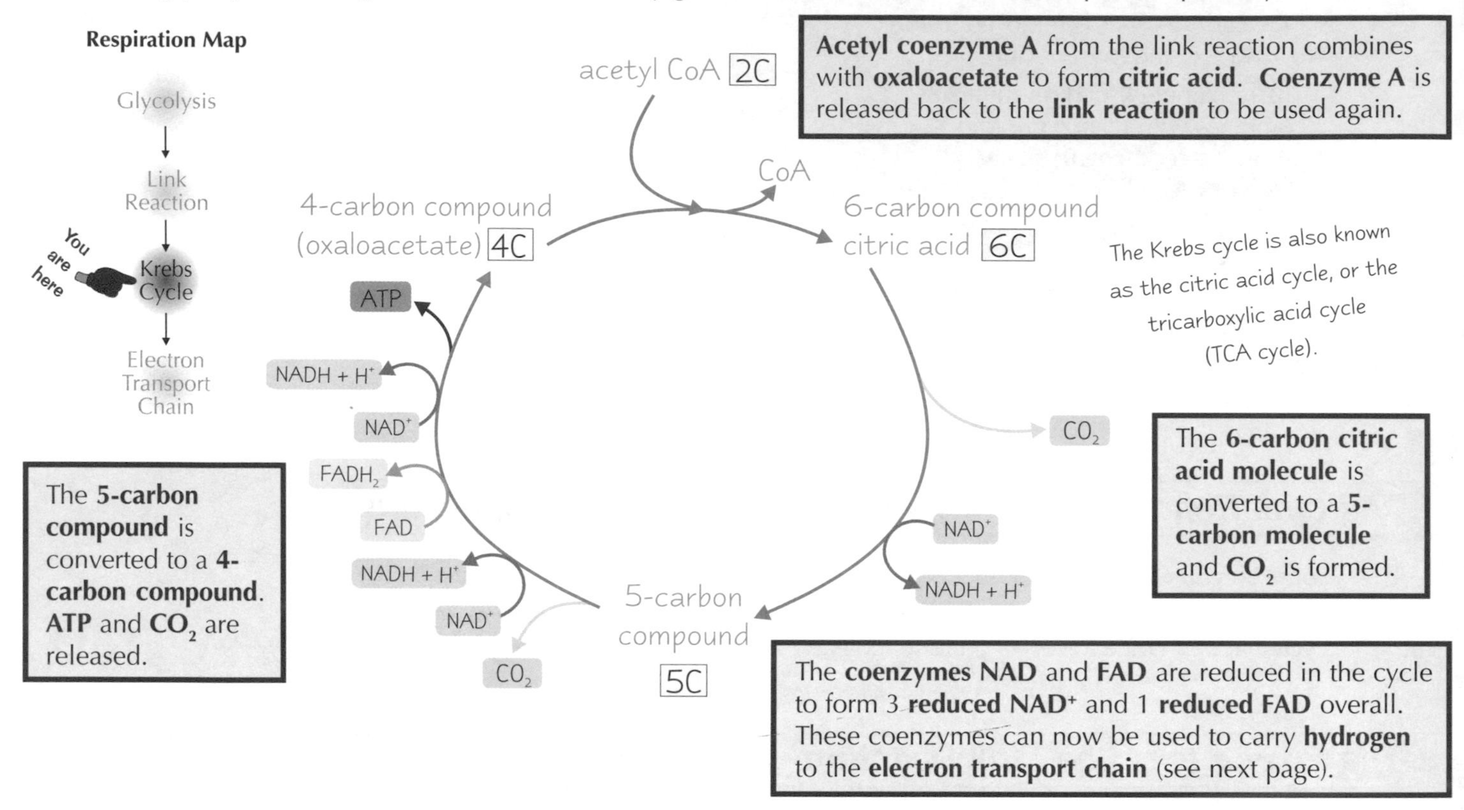

__Products__ of the __Krebs Cycle__ are used in the __Electron Transport Chain__

Some products are **reused**, some are **released** and others are used for the **next stage** of respiration:

- One **CoA** is **reused** in the next **link reaction**.
- **Oxaloacetate** is **regenerated** so it can be **reused** in the next **Krebs cycle**.

- Two **carbon dioxide** molecules are released as a **waste product** of respiration (per turn of the cycle).
- One molecule of **ATP** is made per turn of the cycle — by **substrate level phosphorylation.**

- **Three reduced NAD** and **one reduced FAD** co-enzymes are made (per turn of the cycle) and carried forward to the **electron transport chain**.

*The **Electron Transport Chain** is the **Final Stage** of Aerobic Respiration*

Before we get too bogged down in all the details, here's what the electron transport chain is all about:

All the products from the previous stages are used in this final stage. Its purpose is to **transfer** the **energy** from molecules made in glycolysis, the link reaction and the Krebs cycle to ADP. This forms **ATP**, which can then deliver the energy to parts of the cell that need it.

The electron transport chain is where **most of the ATP** from respiration is produced. In the whole process of aerobic respiration, **32 ATP molecules** are produced from one molecule of glucose: 2 ATP in glycolysis, 2 ATP in the Krebs cycle and 28 ATP in the electron transport chain.

The electron transport chain also **reoxidises NAD and FAD** so they can be reused in the previous steps.

Krebs Cycle and Electron Transport Chain

The *Electron Transport Chain* produces *lots of ATP*

The **electron transport chain** uses the molecules of **reduced NAD** and **reduced FAD** from the previous three stages to produce **28 molecules of ATP** for every molecule of glucose.

1) **Hydrogen atoms** are released from **NADH + H^+** and **$FADH_2$** (as they are oxidised to NAD^+ and FAD). The H atoms **split** to produce **protons (H^+)**, and **electrons (e^-)** for the chain.
2) The **electrons** move along the electron chain (made up of **electron carriers**), losing energy at each level. This energy is used to **pump** the **protons** (H^+) into the space **between** the inner and outer **mitochondrial membranes** (the **intermembranal space**).

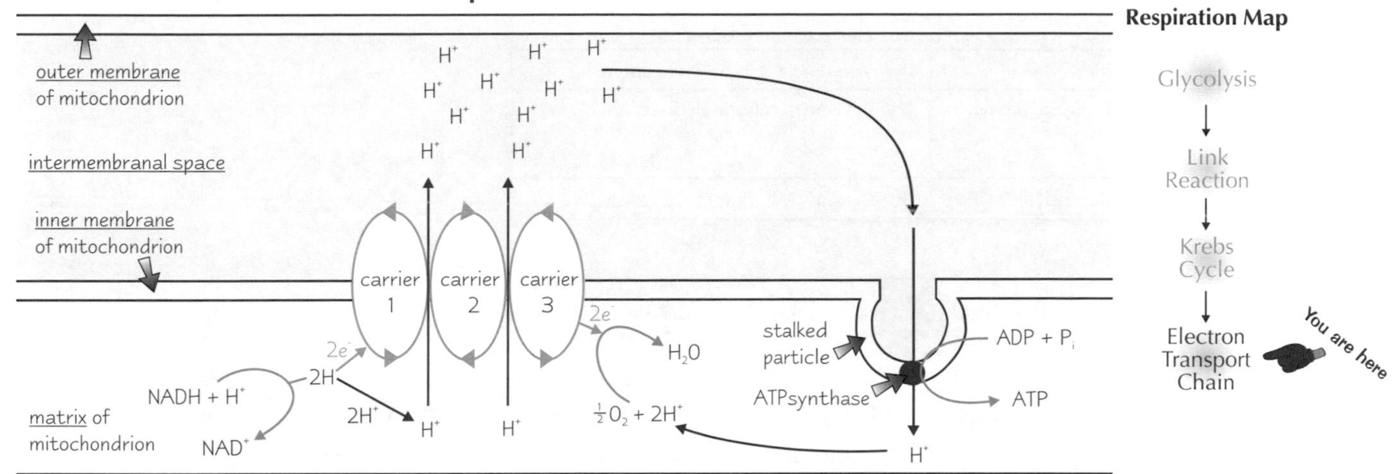

3) The **concentration** of protons is higher in the intermembranal space than in the mitochrondrial matrix, so an **electrochemical gradient** exists.
4) The **protons** then diffuse back through the inner membrane **down** the **electrochemical gradient**, through specific channels on the **stalked particles** of the **cristae** — this drives the enzyme **ATPsynthase.** By 'spinning like a motor', this enzyme supplies **electrical potential energy** to make **ATP** from ADP and inorganic phosphate.
5) The **protons** and **electrons** recombine to form **hydrogen**, and this combines with **molecular oxygen** (from the blood) at the end of the electron transport chain to form **water**. Oxygen is said to be the final **hydrogen acceptor**.

The **synthesis of ATP** as a result of the energy released by the electron transport chain is called **oxidative phosphorylation**.

This is how the **electron transport chain** produces **28** molecules of **ATP** from **1** molecule of **glucose**:

- 1 turn of the Krebs cycle produces **4** molecules of **reduced NAD** (including **1** from the **link reaction**) and **1** of **reduced FAD**.
- **2** molecules of **pyruvate** enter the Krebs cycle for **each** molecule of **glucose**, so overall **8 NAD^+** and **2 FAD** are reduced.
- **2 reduced NAD** are also produced from the first part of respiration, **glycolysis** (see p. 4).
- Each **reduced NAD** can produce **2.5 ATP**, and each **reduced FAD** can produce **1.5 ATP**.
- So: **8** reduced NAD + **2** reduced NAD = **10** reduced NAD. **10 × 2.5 = 25 ATP**. **2 reduced FAD** × **1.5 = 3 ATP**. In total, 25 + 3 = **28 molecules of ATP**.

(There are also **2 ATP** produced by **glycolysis**, and **2** for each molecule of glucose in the **Krebs cycle** = **32 ATP** produced in total by **respiration**.)

Practice Questions

Q1 How many molecules of CO_2 are made in one turn of the Krebs cycle?

Q2 Name the two coenzymes that are reduced in the Krebs cycle.

Q3 Which molecule finally accepts the electrons passed down through the electron transport chain?

Exam Question

Q1 Calculate the number of ATP molecules that are produced by aerobic respiration from one molecule of glucose. Show your working in detail. [14 marks]

Cheers for that Mr Krebs...

...you can keep your cycle, in future. Phew, this biochemistry stuff is tough going. The key to learning this stuff is to learn the big facts first — glycolysis, link reaction, Krebs cycle, electron transport chain. Once you know what the main parts are and roughly what happens at each stage, you stand some chance of learning the more detailed stuff.

Anaerobic Respiration

Anaerobic respiration is respiration without oxygen — so read on — and don't hold your breath.

Anaerobic Respiration is Different from Aerobic Respiration

Anaerobic Respiration	Aerobic Respiration
Does not need oxygen	Needs oxygen
Takes place in cytoplasm	Takes place in cytoplasm and mitochondria of eukaryote cells (or in folded membranes of prokaryote cells)
Uses glycolysis and alcoholic or lactate fermentation	Uses glycolysis, Krebs cycle, link reaction and the electron transport chain
Pyruvate not completely oxidised	Pyruvate oxidised in link reaction
Can follow either of two metabolic pathways 1. alcoholic fermentation 2. lactate fermentation	Follows one metabolic pathway
Produces 2 ATP for every glucose	Produces 32 ATP for every glucose

There are Two forms of Anaerobic Respiration

The **two different kinds** of anaerobic respiration are:

1) **Alcoholic fermentation** — used by **plants** and some **microorganisms** like yeast.
2) **Lactate fermentation** — used by **animals** and some bacteria.

They both occur in the **cytoplasm** and both start with the **glycolysis reaction**.

Alcoholic Fermentation makes Alcohol

Alcoholic fermentation's a bit simpler than aerobic respiration — thank goodness.

1) It starts with **glucose** (**6C**). The process of glycolysis turns this into **pyruvate** (**3C**) (see page 4).
2) **CO_2** is removed from pyruvate to form **ethanal** (**2C**).
3) The **reduced** NAD made in glycolysis is **reoxidised** as **two hydrogen atoms** are accepted by **ethanal** to form **ethanol**.
4) The **reoxidised NAD** can then be reused in **glycolysis**.

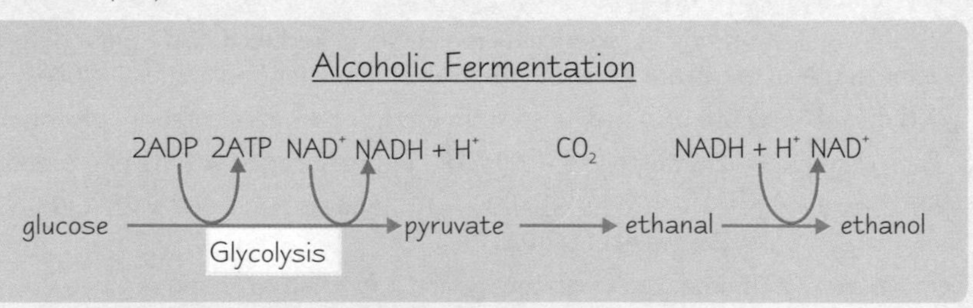

Alcoholic fermentation is used in Industrial Processes

Some industrial processes use fermentation reactions to help produce their products.
The most useful products that **microorganisms** produce by fermentation are **alcohol** and **carbon dioxide**.

Two examples of industries that use fermentation are:
1) **Brewing** — **Yeasts** ferment fruit or grain to produce **alcoholic drinks** like wine, cider and beer.
2) **Breadmaking** — **Yeasts** produce **CO_2** which helps the bread to rise. The alcohol **evaporates** during baking.

Fermentation reactions are normally carried out on an industrial scale in vessels called **fermenters** or **bioreactors** (see pages 34-35 for more).

Anaerobic Respiration

Lactate Fermentation Occurs in Animal Cells and Some Bacteria

Some animal cells can also respire without oxygen when they need to, for a short time:

1) It starts with **glycolysis**, producing **pyruvate** (see page 4).
2) **Pyruvate** is **reduced** to **lactic acid** (lactate) by accepting **two H atoms** from **reduced NAD**.
3) **NAD** is returned to glycolysis to be **used again**.

Pretty simple, really.

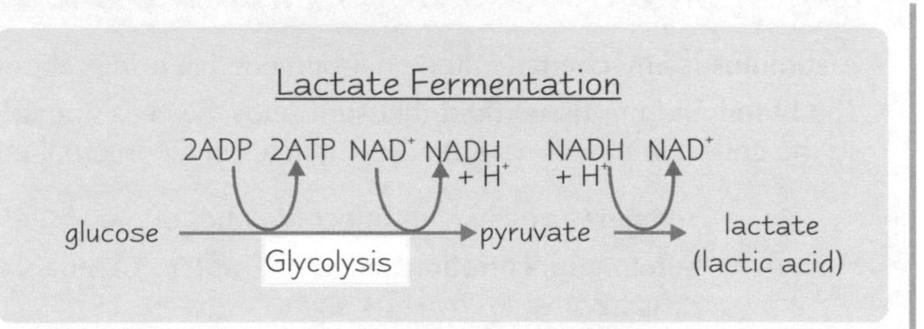

Anaerobic respiration often occurs in Muscle Cells during Vigorous Exercise

When the muscles have been working hard, the supply of pyruvate from glycolysis is **more than** the oxygen supply available for the later stages of **aerobic respiration**. When this happens cells start respiring anaerobically by **lactate fermentation**.

As a result lactate builds up in the muscles. The lactate is removed from the muscle cells by the **blood**. When oxygen is available again, the **liver** converts the lactate back to pyruvate or back to **glucose** and then **glycogen** for storage. The amount of oxygen needed to do this is called the '**oxygen debt**'.

Olympic sprinters run out of oxygen in their muscle cells after about 10 metres of a 100 metre race. From then on they use anaerobic respiration to complete the race.

Anaerobic Respiration is Not very Efficient

In **aerobic** respiration, about **32 ATP** molecules are made for every glucose molecule (see p. 7).

In **anaerobic** respiration, only **2 ATP** molecules are made (during glycolysis).

So **aerobic** respiration releases **loads more energy** than **anaerobic** respiration.

Practice Questions

Q1 Name the two types of anaerobic respiration.

Q2 What molecule is made when CO_2 is removed from pyruvate during alcoholic fermentation?

Q3 What is the final product of alcoholic fermentation?

Q4 Which type of anaerobic respiration occurs in animals?

Q5 Why is anaerobic respiration so inefficient compared to aerobic respiration?

Exam Question

Q1 a) Compare and contrast the two forms of anaerobic respiration. [10 marks]

b) In terms of the amount of ATP produced, how much more efficient is aerobic respiration at releasing energy than anaerobic respiration? [3 marks]

c) Olympic sprinters use anaerobic respiration during most of a 100 m race, even though it is less efficient at releasing energy than aerobic respiration. Suggest why this might be. [2 marks]

Did somebody mention beer...

Anaerobic respiration isn't too bad. It's not overly-complicated, and fermentation even has some rather tasty products. That makes it a lot more interesting than most of this section, I reckon. And everyone knows what lactate fermentation feels like. This is the stuff that makes biology slightly less tedious, so you've got no excuse for not learning this.

Homeostasis and Communication

This section is all about responding to changes in your environment and keeping things constant inside the body. Everything has to be carefully balanced — otherwise your body would be totally out of control.

Organisms need to Detect Internal and External Stimuli

A **stimulus** is any **change** in the environment that brings about a **response** in an organism.

The **blood** and the **tissue fluid** that surrounds the cells (the **internal environment**) needs to be kept within **certain limits**, so the cells can function normally. **Changes** in the internal environment can **damage cells**:

1) **Temperature** changes affect the rates of metabolic reactions and high temperatures can denature proteins.
2) **Solute concentrations** affect the **water potentials** of solutions and therefore the loss or gain of water by cells due to osmosis.
3) Changes in **pH** can affect the function of proteins by changing their tertiary structure (3-D shape).

Changes in the **external environment** also need to be monitored, e.g. in order to detect **food** sources or detect **predators**. If an organism detects these stimuli and responds appropriately then its chances of survival are increased. The body has to respond to changes in the external environment, e.g. if the external temperature is too high the body has to find ways to prevent the internal temperature rising.

Homeostasis keeps the Internal Environment Constant

Homeostasis is the maintenance of a constant internal environment.
A mechanism called **negative feedback** helps to do this:

NEGATIVE FEEDBACK

Changes in the environment trigger a response that **counteracts** the changes — e.g. a **rise** in temperature causes a response that **lowers body temperature**.

This means that the **internal environment** tends to stay around a **norm** (called the **set point**), the level at which the cells work best.

This only works within **certain limits** — if the environment changes too much then the effector may not be able to **counteract** it.

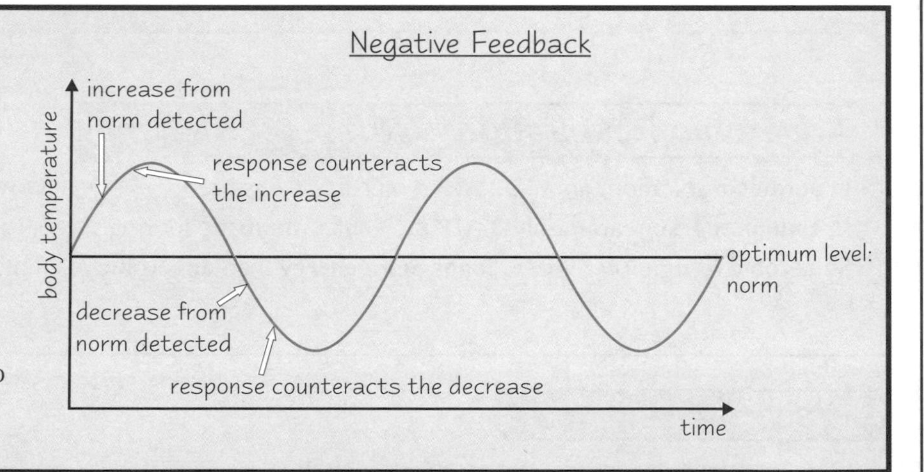

If an organism can regulate its internal environment then it can be **independent** of its external environment, e.g. camels can live in deserts, despite the high temperatures and lack of water because they can regulate their body temperature and water content.

There are lots of examples of negative feedback in this section, e.g the control of blood glucose on pages 16-17 and the regulation of water content on pages 14-15.

The body Responds to internal and external Stimuli in a Series of Stages

1) A **sensory receptor detects** a change (the **stimulus**).
2) The receptor **communicates** with the part of the body that brings about a response (the **effector**), via the **nervous system** or **hormones**.
3) An **effector** brings about the **response**. Glands and muscles are effectors.

This process is the same for responding to external stimuli and internal stimuli. Complex stimuli, e.g. an approaching predator, will involve processing and coordination of the response by the **central nervous system** (see pages 24-25).

Homeostasis and Communication

Organisms have **Sensory Receptors** that are sensitive to **Stimuli**

You can classify receptors in animals depending on the **type of stimulus** they detect:
A **receptor** is the part of the body of an organism that **detects** the stimulus.

1) **Thermoreceptors** are sensitive to **temperature** — they're stimulated by **heat energy**.
2) **Photoreceptors** (e.g. the cells that contain the pigments in the retina of the eye) are sensitive to **light** — they're stimulated by **electromagnetic energy**.
3) **Mechanoreceptors** are sensitive to **sound, touch, pressure** or even **gravity** — they're stimulated by **kinetic energy**.
4) **Chemoreceptors** are sensitive to **chemicals** — they're stimulated by **chemical energy**. They're involved in the senses of smell and taste.

Some kinds of chemicals in **plants**, such as **phytochromes** (see page 81) are receptors — they detect **light**.

The Receptors **Communicate** with **Effectors** to bring about a **Response**

An **effector** is a part of the body that brings about a **response** to the signal from a receptor.
There are two main ways that a receptor **communicates** with an effector:

1) **Chemical Communication**
The receptor produces a **chemical** (hormone) that binds to the effector. The chemical moves from receptor to effector by **diffusion** for short distances, or **mass flow** (transport in bulk) for long distances (see pages 18-19).

2) **Nervous Communication**
The receptor triggers a **nerve** (electrical) **impulse** in nerve cells, which stimulates the effector to respond (see pages 20-21).

The two ways of communication have different characteristics:

	Chemical Communication	Nervous Communication
Speed of transmission	Slow	Fast
Signal carried by	Hormones	Nervous impulses
System used	Endocrine	Nervous
Effects	Widespread	Localised
Length of response	Long	Short

In animals, effectors are usually **muscles** (the response is **contraction**), or **glands** (the response is **secretion**). See pages 58-59 for more about muscles.

Practice Questions

Q1 Why do you need to detect changes in your environment?

Q2 Define homeostasis.

Q3 Give one factor that is controlled by homeostasis in mammals, and explain why it needs to be controlled.

Q4 Explain how stimulus detection results in a response.

Q5 Name two types of sensory receptor.

Exam Questions

Q1 a) Explain what is meant by the term 'negative feedback'. [2 marks]
b) Give two examples of factors that are controlled by a negative feedback mechanism. [2 marks]

Q2 a) Outline the two main ways in which the receptors of the nervous system communicate with effectors. [3 marks]
b) Give two differences between these two types of communication. [2 marks]

My biology teacher often gave me negative feedback...

The key to understanding homeostasis is getting your head around negative feedback. It's not complicated — if one thing goes up, the body responds to bring it back down, or vice versa. Look at pages 14-17 for more negative feedback loops.

The Kidneys and Excretion

Here's two lovely pages for you to get stuck into on how the kidneys excrete stuff our bodies don't need. Then afterwards you can get stuck into a big steak and kidney pie.

*The **Kidneys Excrete Urea** made by the **Liver***

Excess **amino acids** in the body (from **protein** in the diet) can't be stored for use later on, so they are converted to urea in the liver:

1) Nitrogen-containing **amino groups** from the excess amino acids are removed, forming **ammonia** and **organic acids**. This is **deamination**.
2) Ammonia is too poisonous for mammals to excrete directly, so it's combined with CO_2 — converting it to a safer form — **urea**.
3) The urea is **released** into the **blood**.

Urea and other waste substances produced by the liver are **excreted** from the body by the **kidneys**.

When the blood passes through the kidney nephrons (the tubes that run through the kidneys — see below for more on these), liquid is filtered out of the blood, carrying small solutes with it, including **urea**.

The useful solutes are reabsorbed, and the waste products are removed from the body in **urine** (see opposite page).

You need to know about the **two main stages** of kidney function — **ultrafiltration** in the renal capsules (also called the Bowman's capsules) and **selective reabsorption** in the medulla.

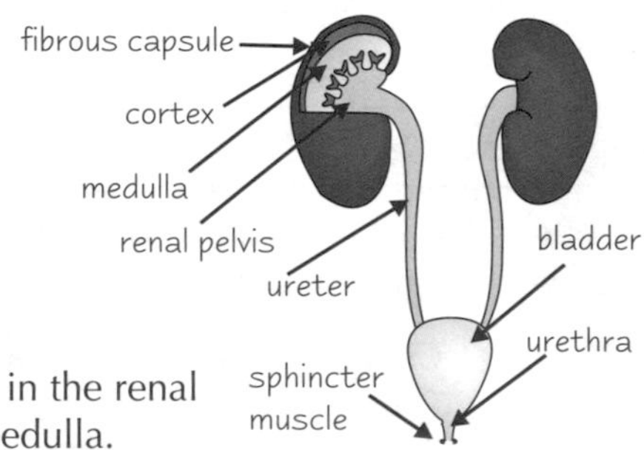

Ultrafiltration** in the kidneys takes place in the **Renal Capsules

Blood enters the kidney cortex through the **renal artery** then goes through millions of knots of capillaries in the kidney cortex. Each knot (**glomerulus**) is a bundle of capillaries looped inside a hollow ball called a **renal** (or Bowman's) **capsule**. An **afferent arteriole** takes blood into each glomerulus and an **efferent arteriole** takes blood out.

Ultrafiltration takes place in glomeruli. Blood pressure squeezes liquid from the blood through the capillary wall. Small molecules and ions pass through but larger ones like proteins and blood cells stay behind in the blood. The liquid collects in microscopic **tubules**.

Useful substances are **reabsorbed** back into the blood, and waste substances (like dissolved urea) are excreted in the urine.

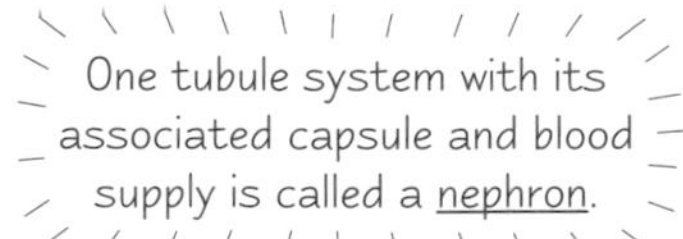

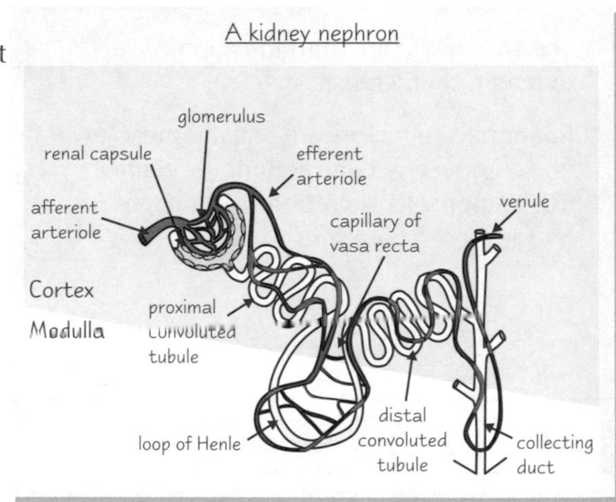

Ultrafiltration in the capillary and renal capsule membranes

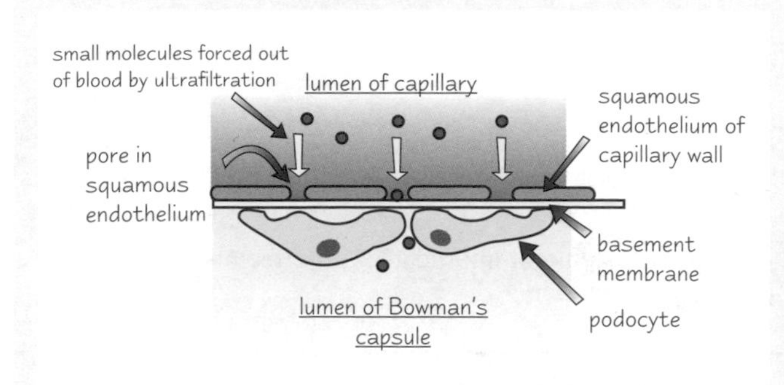

Small molecules and ions pass from blood in the glomerular capillary into the renal capsule through:

1) Pores in the **capillary wall** (the wall is made of one layer of flat cells — squamous endothelium).
2) A **basement membrane** made up of collagen fibres and glycoprotein. This membrane stops large molecules passing through, so acts as a **filter**.
3) A specialised epithelium of the renal capsule, made up of cells called **podocytes**. These support the membrane while letting the filtrate pass through.

The Kidneys and Excretion

Useful substances are Reabsorbed

Filtrate from the renal capsule enters the **proximal convoluted tubule** in the cortex of the kidney. The wall of this tubule is made of **epithelial cells**, with **microvilli** facing the filtrate to increase the surface area of the epithelial cells.

Blood leaving the glomerulus along the efferent arteriole enters another capillary network, called the **vasa recta**, that's wrapped around the proximal convoluted tubule. This provides a large surface area for **reabsorption** of useful materials from the **filtrate** (in the tubules) into the **blood** (in the capillaries) by:

1) **active transport** of glucose, amino acids, vitamins and some salts
2) **osmosis** of water

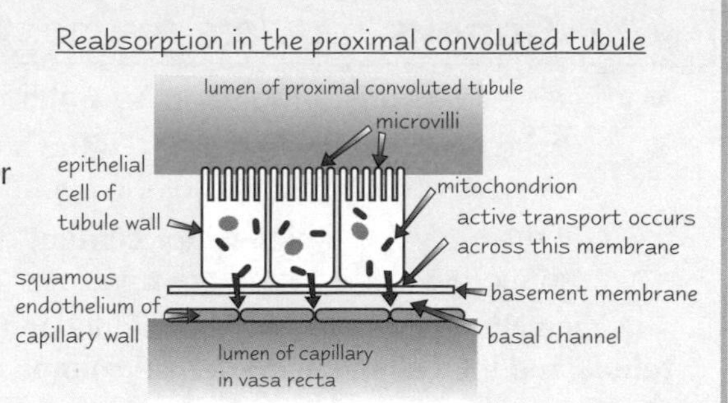

Water is reabsorbed from the Whole Tubule

Water enters the blood by **osmosis** because the water potential of the blood is **lower** than that of the filtrate. Most of the water is reabsorbed from the **proximal convoluted tubule**. Reabsorption from the **distal convoluted tubule** and **collecting duct** is controlled by **hormones** (see pages 14-15 for more on this).

Urine is a Mixture of substances dissolved in Water

Urine usually contains the following things:

1) **Variable amounts** of **water** and **dissolved salts**, depending upon much you've drunk (see pages 14-15).
2) **Variable amounts** of **dissolved urea**, depending upon how much protein you've eaten.
3) Other substances such as hormones and water-soluble vitamins.

It **doesn't** usually contain:

1) **Proteins**, because they're too big to be filtered out in the renal capsule (they can't pass through the basement membrane).
2) **Glucose, amino acids** or **fat-soluble vitamins**, because they're actively reabsorbed back into the blood from the proximal convoluted tubule.
3) **Blood cells**.

Don't get mixed up — urea is a specific kind of substance, but urine is a mixture that contains urea.

Practice Questions

Q1 Describe the process of ultrafiltration.

Q2 Name three components of the filtrate of a nephron that will be reabsorbed back into the blood from the proximal convoluted tubule.

Q3 What are the names of the processes that account for the reabsorption of substances?

Exam Questions

Q1 Explain why the urea concentration in urine will rise after consumption of a meal that is rich in protein. [6 marks]

Q2 a) Suggest how the features of the proximal convoluted tubule of a kidney nephron maximise the rate of absorption of glucose. [5 marks]

b) Suggest why a smaller quantity of urea passes from the tubule into the blood than glucose. [3 marks]

Q3 Occasionally some people produce urine that contains traces of protein. Suggest an explanation for this. [2 marks]

It's steak and excretion organ pie for dinner...

The kidneys are pretty complicated organs. That's why it's so serious when they go wrong — all that toxic urea would just stay in your blood and poison you. If your kidneys fail you'll end up hooked up to a machine for hours every week so it can filter your blood, unless some kind person donates a new kidney for you.

The Kidneys and Osmoregulation

You should have a rough idea about how the kidneys work by now, so here's a bit more detail on how they help with homeostasis. Read. Learn. Enjoy.

The **Kidneys** regulate the body's **Water Content (Osmosregulation)**

Mammals excrete urea in solution, so **water** is lost too. The kidneys regulate the levels of water in the body.

1) If the body is **dehydrated** (e.g. if the body has lost a lot of water by sweating) then more water is **reabsorbed** by osmosis from the tubules of the nephron, so less water is lost in the urine.
2) If the body has a **high water content** (e.g. from drinking a lot) then **less** water is reabsorbed from the tubules, so more water is lost in the urine.

This regulation takes place in the middle and last parts of the nephron — the **loop of Henle**, the **distal convoluted tubule** and the **collecting duct**. The **volume** of water reabsorbed is controlled by hormones.

The **Loop of Henle** has a **Countercurrent Multiplier Mechanism**

This sounds really scary, but stick with it, it's not too complicated really:

Just between the proximal and distal convoluted tubules is a part of the nephron called the **loop of Henle**. It's made up of two 'limbs'.

1) Water leaves the **descending limb** of the loop by osmosis, and Na^+ and Cl^- ions diffuse into the loop. This makes the **ion concentration** of the tubule **higher** towards the **base** of the loop.
2) The Na^+ and Cl^- ions are then actively pumped out of the top of the **ascending limb** of the loop into the medulla. The high concentration of Na^+ and Cl^- ions in the medulla causes water to leave the collecting duct and descending limb by **osmosis**.
3) The water is then **reabsorbed** into the blood through the **capillary network**.

This mechanism is called the **countercurrent multiplier**.

The countercurrent multiplier mechanism lets land-living mammals produce urine with a **solute concentration** higher than that of the blood, so they can avoid losing too much water. The volume of water reabsorbed can be **regulated** depending on the needs of the body.

Another hard day studying reading, writing and the countercurrent multiplier mechanism.

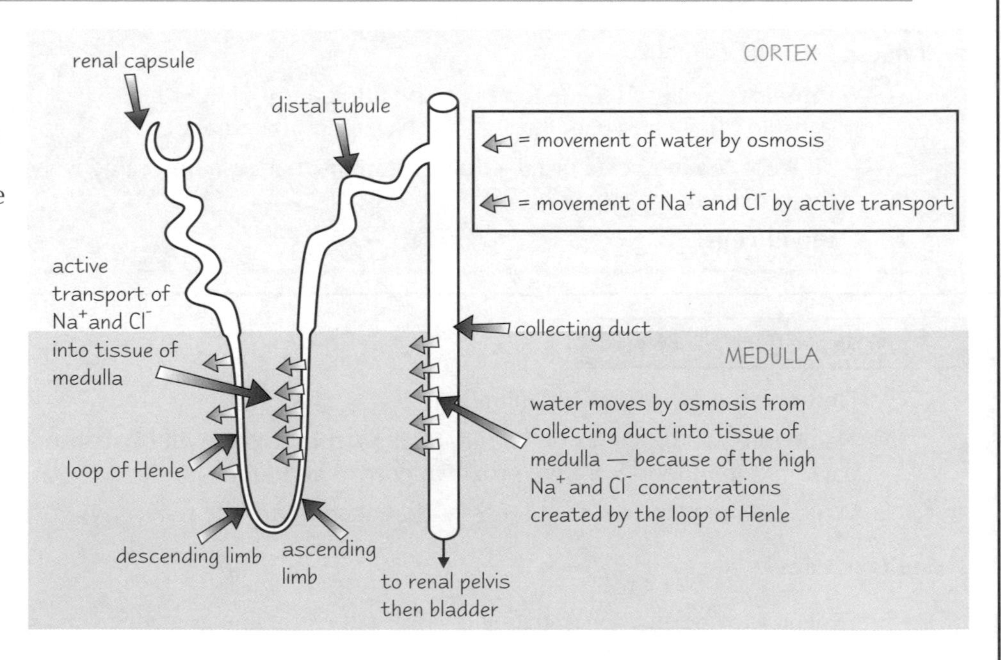

Water Reabsorption is controlled by **Hormones**

A hormone released from the **posterior pituitary gland**, called **antidiuretic hormone (ADH)**, makes the walls of the distal convoluted tubules and collecting ducts **more permeable** to **water**. More water is then **reabsorbed** from these tubules into the medulla (and then into the blood) by **osmosis**. This means that **less water** is lost in the urine.

It's called antidiuretic hormone because diuresis is when lots of dilute urine is produced.

The Kidneys and Osmoregulation

Blood ADH levels are High when you're Dehydrated

Dehydration is what happens when you **lose water**, e.g. by sweating during exercise:

1) The water content of the blood drops, so its **water potential drops**.
2) This is detected by **osmoreceptors** in the **hypothalamus**.
3) This stimulates the **posterior pituitary gland** to release **more ADH** into the blood.
4) The ADH **increases** the **permeability** of the walls of the **collecting ducts** in the kidneys, so **more water** is **reabsorbed** back into the blood by **osmosis**.
5) **Less water** is lost in the urine.

Blood ADH levels are Low when your body is Hydrated

If you drink lots of water, more water is absorbed from the gut into the **blood**, and the **excess** is lost in the **urine**:

1) The water content of the blood rises, so its **water potential rises**.
2) This is detected by the **osmoreceptors** in the **hypothalamus**.
3) This stimulates the **posterior pituitary gland** to release **less ADH** into the blood.
4) Less ADH means that the collecting ducts are less permeable, so **less water** is **reabsorbed** into the blood by **osmosis**.
5) **More water** is lost in the urine.

Osmoregulation is an example of Negative Feedback

Here's a summary diagram to show how osmoregulation maintains the internal environment by negative feedback:

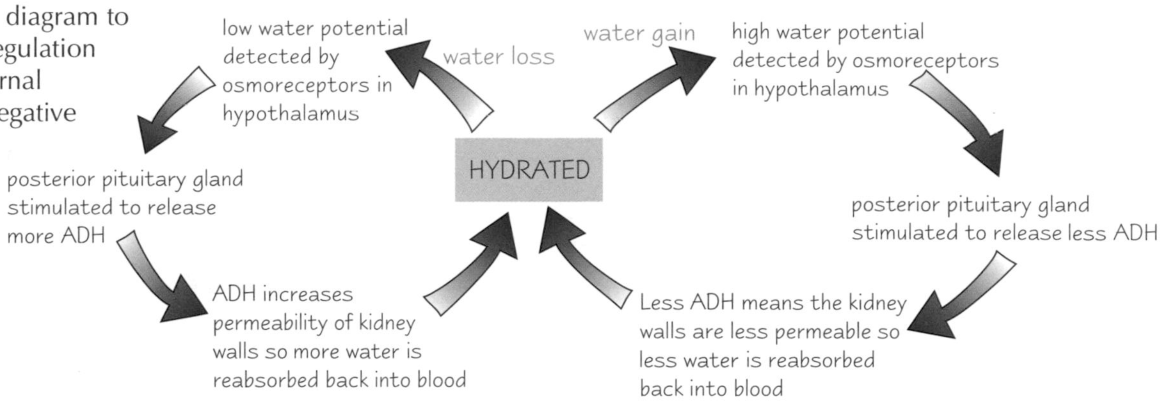

HYDRATED
water loss
low water potential detected by osmoreceptors in hypothalamus
posterior pituitary gland stimulated to release more ADH
ADH increases permeability of kidney walls so more water is reabsorbed back into blood
water gain
high water potential detected by osmoreceptors in hypothalamus
posterior pituitary gland stimulated to release less ADH
Less ADH means the kidney walls are less permeable so less water is reabsorbed back into blood

Practice Questions

Q1 What are the main ways in which water can be lost from the body of a terrestrial animal?
Q2 Describe how the loop of Henle increases the salt concentration of the medulla of the kidney.
Q3 Explain the importance of the medulla in allowing reabsorption of water from urine.
Q4 Which gland releases ADH?
Q5 What is the effect of ADH on kidney function?

Exam Question

Q1 Levels of ADH in the blood rise during strenuous exercise.
Explain the cause of the increase and the effects it has on kidney function. [10 marks]

If you don't understand what ADH does, urine trouble...

Seriously, though, there are two main things to learn from these pages — the countercurrent multiplier mechanism and the role of ADH in controlling the water content of urine. You'll need to be able to identify the different parts of the kidney nephron too. Keep writing it down until you've got it sorted in your head and you'll be just fine.

Blood Glucose Control

Ever felt a 'sugar rush' after eating eighteen packets of Refreshers? Ok, maybe no one else eats that many at once. These two pages are about how your body deals with all the glucose you get from your food.

Many Factors can change the **Blood Glucose Concentration**

Glucose enters the blood from the **small intestine** and dissolves in the blood plasma. Its concentration usually stays around **100mg per 100cm³ of blood**, but some factors can change this level:

1) Blood glucose concentration **increases** after consuming food, especially if the food is **high** in carbohydrate.
2) Blood glucose concentration **falls** after exercise, because more glucose is used in respiration to release energy.

Big changes in the concentration of blood glucose can **damage cells** by changing their water potential.

The **Pancreas** secretes **Hormones** to control **Blood Glucose Levels**

Two hormones — **insulin** and **glucagon**, regulate blood glucose concentration. They're secreted by clusters of cells in the **pancreas** called the **Islets of Langerhans**. These cells detect changes in blood glucose concentration.

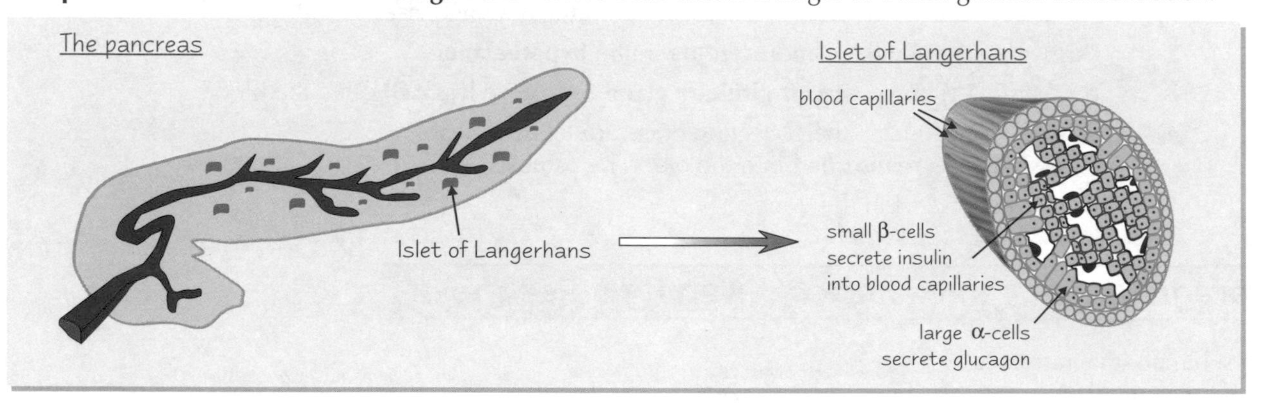

Controlling **Blood Glucose Levels** is another example of **Negative Feedback**

Insulin and glucagon regulate the blood glucose concentration by **negative feedback** (page 10):

When there's a rise in blood glucose concentration...

1) The glucose molecules bind to receptors in the cell membranes of small **beta (β) cells** in the Islets of Langerhans.
2) These cells then secrete **insulin** directly into the blood.
3) Insulin molecules bind to receptors in the cell membranes of **hepatocytes** (liver cells) and other cells, e.g. muscle cells.
4) This **increases** the **permeability** of the hepatocyte cell membranes to glucose, so more glucose is absorbed.
5) Inside the hepatocytes, the insulin activates an **enzyme** that catalyses the condensation of glucose molecules into **glycogen**, which is stored in the cytoplasm of the hepatocytes and in the muscles. This process is called **glycogenesis**.
6) Insulin also increases the rate of respiration of glucose in other cells. The blood glucose concentration **decreases**.

When there's a fall in blood glucose concentration...

1) The larger **alpha (α) cells** of the Islets of Langerhans secrete the hormone **glucagon** directly into the blood.
2) Glucagon binds to receptors on the **hepatocytes**.
3) This activates an **enzyme** inside the hepatocytes that catalyses the **hydrolysis** of stored glycogen into **glucose**. This process is called **glycogenolysis**.
4) The blood glucose concentration **increases**.

Blood Glucose Control

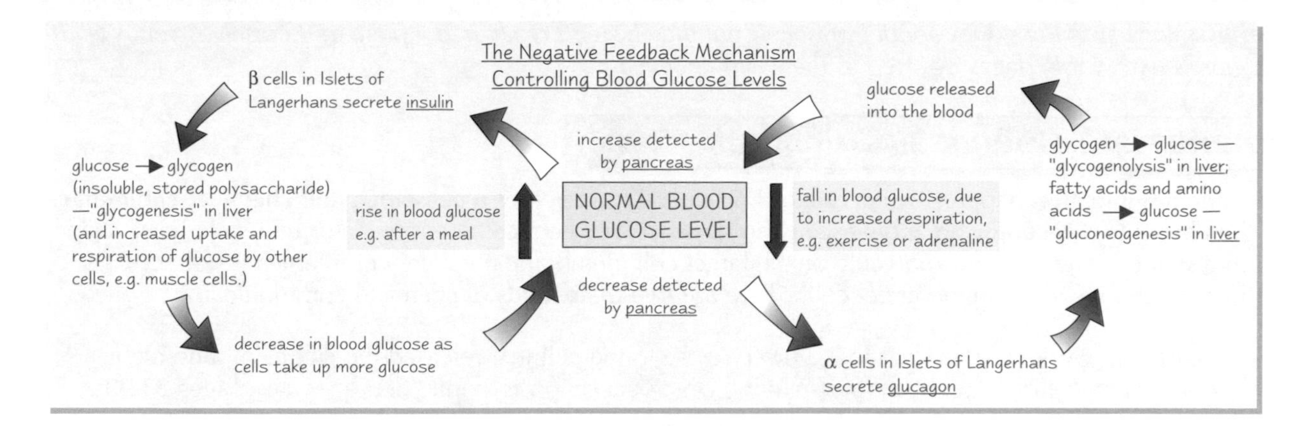

Other Hormones can affect *Blood Glucose Level* too

Other hormones can affect glucose levels in similar ways, although they bind to different receptors. For example, **adrenaline** is secreted during exercise (during a 'fight or flight response') and activates enzymes that **hydrolyse** stored glycogen into glucose, ready for increased muscle activity. It also increases the rate of respiration taking place inside the cells.

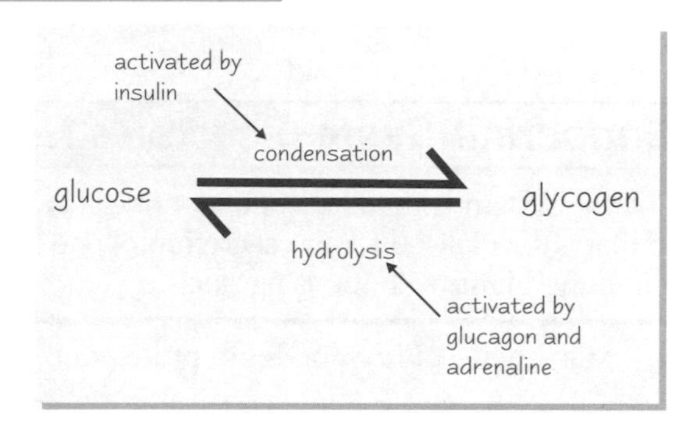

The *Liver Metabolises* glucose and glycogen

Insulin and glucagon control the level of blood glucose but it is the **liver** that actually **removes** the glucose from the blood (forming glycogen). The liver also breaks down stored substances to **release** glucose into the blood. This occurs in the **hepatocytes** (liver cells).

When there's a **shortage** of blood **glucose**, the **liver** produces glucose from:

1) **Glycogen break down**.
2) Other sugars (e.g. **fructose**).
3) **Lipids** (via **glycogen**).
4) **Glycerol** (part of a lipid molecule) and **amino acids**.

Practice Questions

Q1 Give one factor that will increase blood glucose concentration and one that will reduce it.

Q2 What are the roles of the alpha and beta cells of the Islets of Langerhans?

Q3 State two effects of insulin on liver cells.

Q4 Name two hormones that increase the blood glucose concentration.

Q5 Describe how the liver helps regulate blood glucose levels.

Exam Questions

Q1 Explain the role of the endocrine system in returning the blood glucose level to normal after consuming a meal that has a high glucose content. [10 marks]

Q2 Explain why glucagon levels in the blood increase during exercise. [5 marks]

My glucose levels are low — pass the chocolate...

Learn this carefully or you'll end up getting your hormones confused — insulin causes a reduction in blood glucose and glucagon increases it. And make sure you're clear on where the different hormones are produced and where they act. Now eat a huge meal, wait for your blood glucose levels to rise, then draw the diagram from this page until you know it.

Chemical Communication

All this physiology stuff talks a lot about hormones, but they haven't really had a page to themselves yet. Well here you go, here's two whole pages devoted to the weird chemicals...

Hormones are *Chemical Communicators*

Chemical communication happens when one cell produces and releases a substance (the **chemical communicator**), which then binds to a **receptor** on another cell (the **target**). The chemical communicator molecule has a shape that fits into the shape of the receptor molecule on the target cell membrane (they are complementary shapes). This brings about a response in the target cell. There are **two main kinds** of chemical communicator:

1) Those that work only on cells that are very close to the cell that released the chemical, and often move by **diffusion**. E.g. **neurotransmitters**, which act on postsynaptic neurones (see pages 22-23).
2) **Hormones** — these work on targets that are much further away so they enter the blood and are carried by **mass flow** due to the pumping action of the heart.

*The **Endocrine System** secretes **Hormones***

The endocrine system is made up of endocrine glands that secrete chemicals called hormones. A **gland** is any structure that is specialised for the **secretion** of one or more types of substance. Endocrine glands secrete **hormones directly** into the **blood** without using ducts.

1) Many types of hormones are **proteins** or smaller **peptides**, e.g. **insulin** and **adrenaline**.
2) Other hormones are fatty **steroids**, e.g. **oestrogen**, **progesterone** and **testosterone**.

*Hormones are secreted into the **Blood***

1) Hormones are secreted when the **endocrine gland** is **stimulated**. Some glands are stimulated by a change in concentration of a specific solute (sometimes another type of hormone). Others are stimulated by **action potentials** arriving from the nervous system.
2) The hormone is **secreted** and **diffuses** into **blood capillaries**.
3) The hormone is circulated around the body by **mass flow** through the **bloodstream**.
4) The hormone diffuses out of the blood capillaries at different parts of the body. However, it will only bind to **cell-surface receptors** with complementary-shaped **binding sites**. This means that only **target cells** with the correct receptors will respond. Very **small** concentrations of hormones are needed to give a response in target cells. This response usually activates **enzymes** inside the cells.

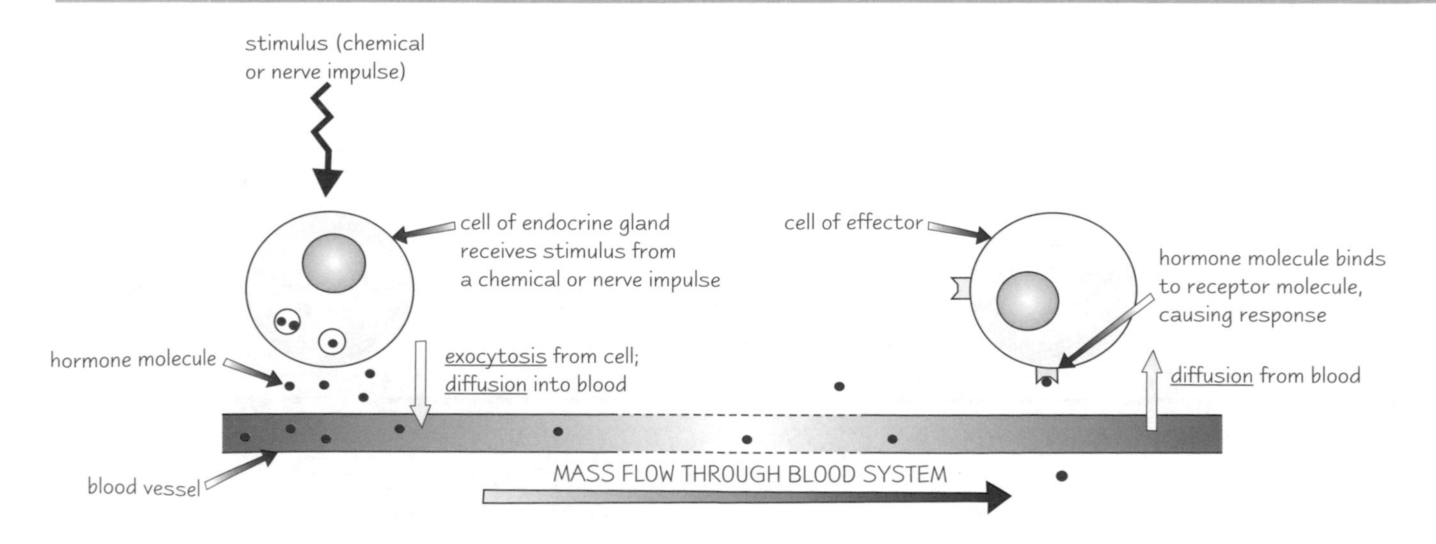

Chemical Communication

Hormonal Control is often by Negative Feedback

As you've seen, hormones that **regulate** factors in the body, such as **glucose concentration**, often work by **negative feedback** (see pages 10-11). This happens when a change in a factor is detected and the response **counteracts** the change.

Negative feedback control happens when hormones are working in **pairs**, e.g. **insulin** and **glucagon** (pages 16-17). Insulin reduces the glucose levels and glucagon raises them. Pairs of hormones that work like this are said to be **antagonistic**.

One final *Negative Feedback* example — *Reproductive Hormones*

The stages of the human **menstrual cycle** are carefully controlled by four hormones and **negative feedback loops** exist to keep the system in balance.

The four hormones are:

1) **Follicle Stimulating Hormone (FSH)**
2) **Luteinising Hormone (LH)**
3) **Oestrogen**
4) **Progesterone**

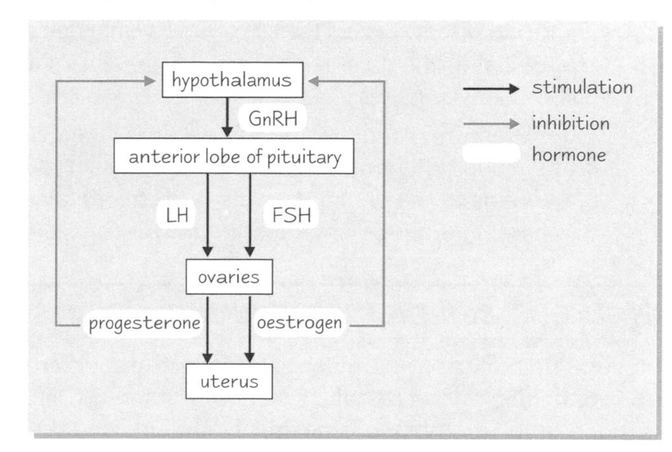

The **hypothalamus** controls the levels of these hormones by negative feedback. It releases the hormone **GnRH** (gonadatrophin releasing hormone), which stimulates the pituitary gland to release **LH** and **FSH**. LH and FSH then stimulate the release of **oestrogen** and **progesterone**. But when levels of oestrogen and progesterone get **too high**, they **inhibit** the release of GnRH from the hypothalamus, so the release of **all the others** stops too. Then when levels **fall** again, the hypothalamus stops being inhibited and everything gets going again.

Practice Questions

Q1 Explain what is meant by the term endocrine gland.

Q2 Give two examples of hormones that are proteins and two that are steroids.

Q3 Explain how a hormone reaches its target cell.

Q4 Name the two hormones that control blood glucose concentration.

Q5 Which endocrine gland releases ADH?

Q6 What does ADH control?

Q7 Name four hormones involved in the human menstrual cycle.

Exam Questions

Q1 Explain why hormonal control in the body brings about slow responses. [1 mark]

Q2 Explain what happens after an endocrine gland is stimulated to produce a hormone. [5 marks]

Bet you never knew you had so many hormones...

Or maybe you did know that. In fact maybe you know everything in this section already, and you're just reading it so that you can chuckle secretly to yourself as you watch lesser mortals trying desperately to learn information that you've known instinctively since you were born. If so, you should probably be writing this book.

Nervous Communication

It's no good being aware of your environment if you can't do anything about it. That's where neurones come in.

Nerve Cells have Polarised Membranes so they can carry Electrical Signals

1) The **nervous system** is made up of nerve cells called **neurones**. Each neurone consists of a **cell body** and extending **nerve fibres**, which are very thin cylinders of cytoplasm bound by a cell membrane. Neurones carry waves of electrical activity called **action potentials** (nerve impulses). They can carry these impulses because their cell membranes are **polarised** (see below) — there are different **charges** on the inside and outside of the membrane.
2) The nerve fibres let the neurones carry action potentials over **long distances**. There are tiny gaps, called **synapses**, between the different nerve fibres. Action potentials can't cross, so a chemical called a **neurotransmitter** is secreted at the synaptic knob to diffuse across the gap. This stimulates a **new** action potential in the next nerve fibre on the other side of the synapse (see pages 22-23).

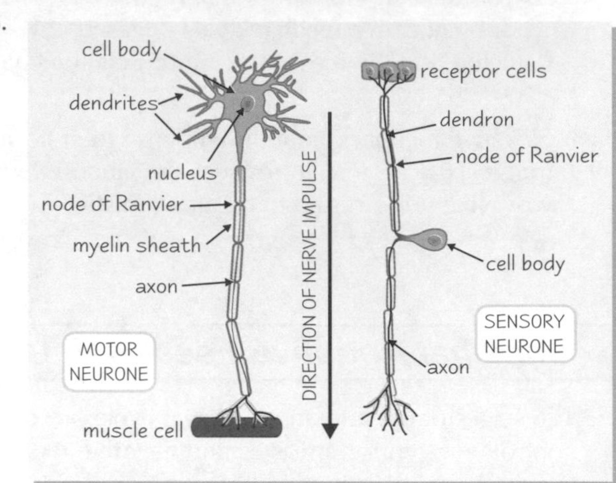

3) Between the receptors and the effectors, the **central nervous system** (i.e. brain and spinal cord) **coordinates** the action potentials passing through the nervous system. **Sensory neurones** carry action potentials from receptors to the central nervous system. **Relay neurones** carry action potentials through the central nervous system, and **motor (effector) neurones** carry them from the central nervous system to effectors.

Neurone cell membranes are Polarised when they're Resting

Resting neurones have a **potential difference** (a difference in **charge**) of about **-65 millivolts** (mV) across their cell membranes. This is because the **outer** surface of the membrane is **positively** charged and the **inner** surface **negatively** charged — the -65 mV is the **overall difference** in charge between them. This is the **resting potential** of the membrane, which is said to be **polarised**.

The resting potential is generated by **sodium-potassium pumps** and **potassium channels** in the membrane. Each sodium-potassium pump moves **three sodium ions** out of the cell by **active transport** for every **two potassium ions** it brings in. The potassium channels then allow **facilitated diffusion** of potassium ions back out of the cell. The inner surface of the membrane becomes more negative than the outer surface because overall, more positive ions move **out** of the cell than move **in**.

Sodium-potassium pump and potassium channel

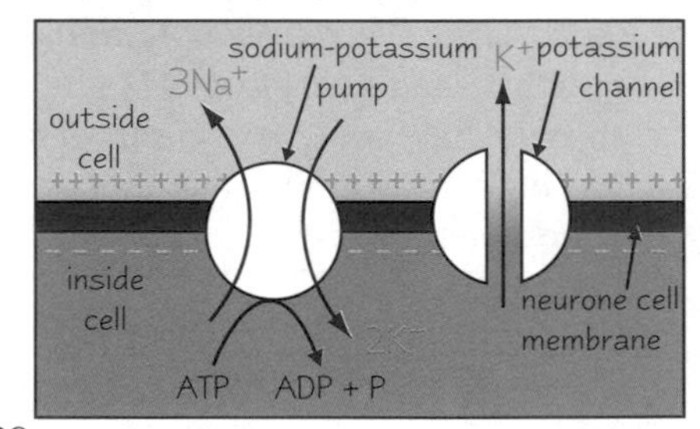

Neurone cells become Depolarised when they're Stimulated

The **sodium-potassium pumps** work pretty much all the time, but **channel proteins** (like the potassium channels) can be opened or closed. **Depolarisation** of neurone cell membranes involves another type of channel protein, **sodium channels**. If a neurone cell membrane is stimulated, sodium channels **open** and **sodium ions** diffuse in. This **increases** the positive charge **inside** the cell, so the charge across the membrane is **reversed**. The membrane now carries a potential difference of about **+40 mV**. This is the **action potential** and the membrane is **depolarised**.

When sodium ions diffuse into the cell, this stimulates nearby bits of membrane and **more** sodium channels open. Once they've opened the channels automatically **recover** and close again.

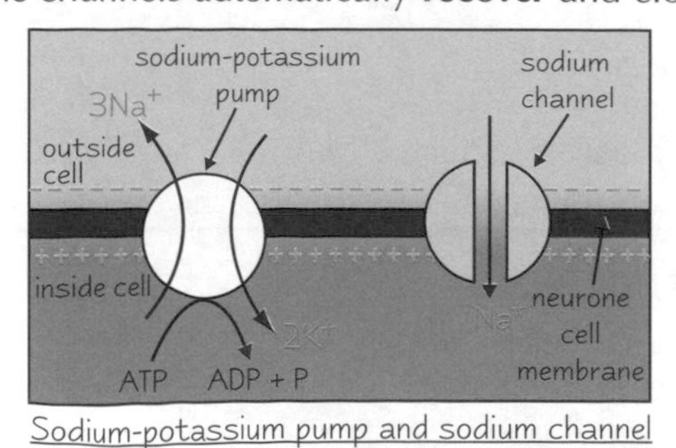

Sodium-potassium pump and sodium channel

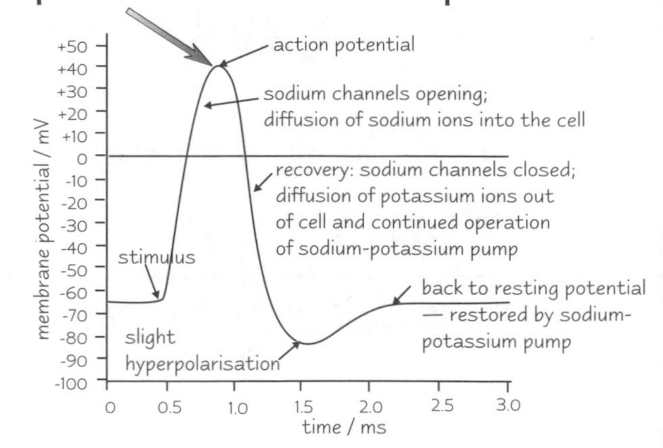

Nervous Communication

*Remember these important features of **Action Potentials***

1) **Nerve axons** in vertebrates are usually covered in a layer of **myelin sheath**, which is produced by **Schwann cells**. Myelin is an **electrical insulator**. Between the sheaths there are tiny patches of bare membrane called **nodes of Ranvier**, where sodium channels are **concentrated**. Action potentials **jump** from one node to another, which lets them move **faster** (this is called **saltatory conduction**).

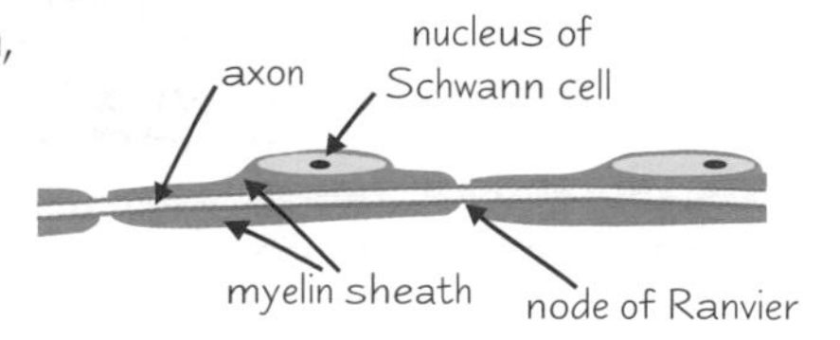

2) Action potentials also go faster along axons with **bigger diameters**, because there's less **electrical resistance**.

3) They go faster as **temperature** increases too, up to around **40°C**. After that, the proteins begin to **denature**.

4) Action potentials have an **all-or-nothing** nature. A stimulus of a certain strength (a **threshold stimulus**) must be applied to get an action potential.

5) A **stronger stimulus** just increases the **frequency** of the action potentials. The **size** of the action potentials stays the same.

6) Straight after an action potential has been generated, the membrane enters a short **refractory period** when it can't be stimulated, because the sodium channels are **recovering** and can't be opened. This makes the action potentials pass along as **separate signals**.

7) Action potentials are **unidirectional** — they can only pass in one direction.

Practice Questions

Q1 What do sensory, relay and motor neurones do in the nervous system?

Q2 What happens to sodium ions when a neurone membrane is stimulated?

Q3 Give two factors that increase the speed of conduction of action potentials.

Q4 What is meant by the 'all-or-nothing' nature of action potentials?

Exam Questions

Q1 The graph shows an action potential 'spike' across an axon membrane following the application of a stimulus.

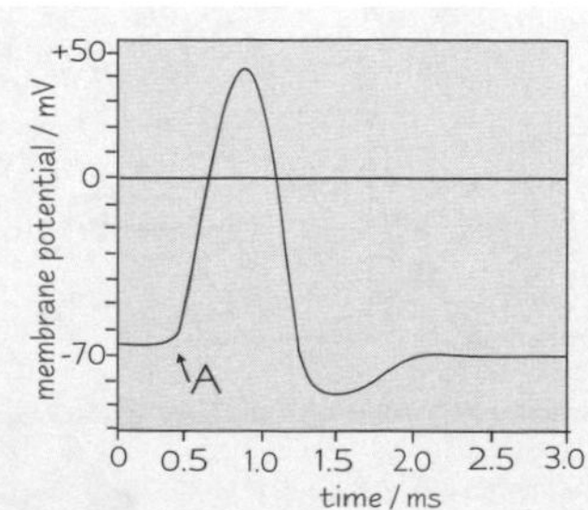

a) What label should be added at point A? [1 mark]

b) Explain what causes the change in potential difference from -65 to +40 mV. [3 marks]

c) Another stimulus was applied at 1.5 ms, but failed to produce an action potential. Suggest why. [2 marks]

Q2 Multiple sclerosis is a disease of the nervous system characterised by damaged myelin sheath. Suggest and explain how this will affect the transmission of action potentials. [5 marks]

I'm feeling a bit depolarised after all that...

The nervous system can seem like a really hard subject at first, but once you've gone over it a couple of times it starts to make sense. Nerves work because there's a charge across their membranes and it's a change in this charge that sends the impulse along the nerve. The charge is set up using ions, which can then move in and out to change the charge.

Synapses

This page is all about synapses, which are the little gaps between the end of one neurone and the start of the next one. Seems like quite an insignificant little thing to fill a whole two pages with, but never mind.

There are **Gaps** between **Neurones**

A **synapse** is a gap between the end of one **neurone** and the start of the next. The main function of a synapse is to **transmit information between neurones**.

An action potential arrives at the end of the axon of the **presynaptic neurone** (the neurone before the synapse), where there's a swelling called a **bouton** or **synaptic knob**. This has **vesicles** containing a chemical **neurotransmitter**.

The impulse passes across the synapse as follows:

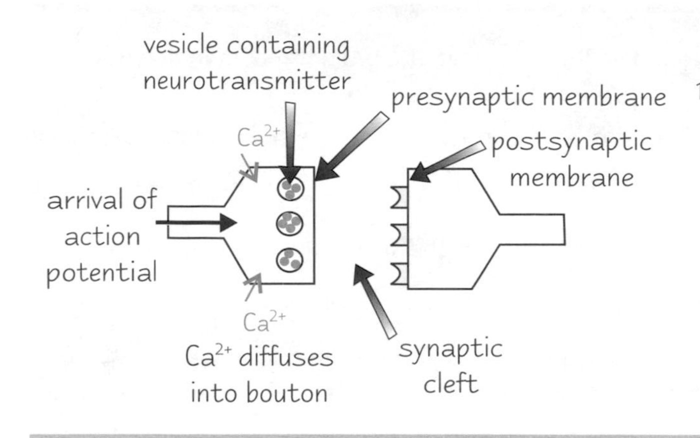

1) The action potential opens **calcium channels** in the membrane, allowing calcium ions to diffuse **into** the bouton. Afterwards these are pumped back out using ATP.

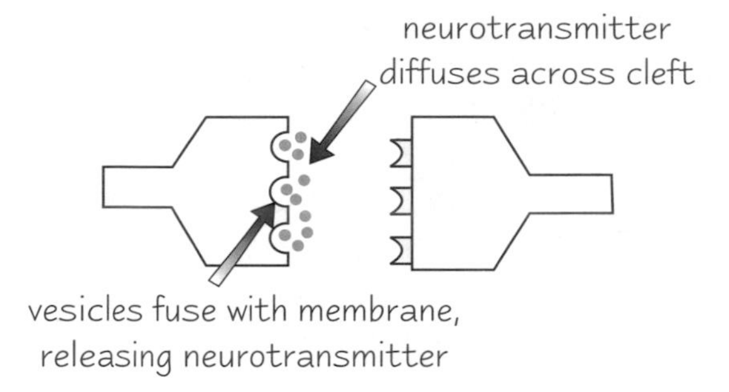

2) The increased concentration of calcium ions in the bouton causes the **vesicles** containing the **neurotransmitter** to move up to and to fuse with the **presynaptic membrane**. This also requires ATP (it's an active process).

3) The vesicles **release** their neurotransmitter into the **synaptic cleft** (this is called **exocytosis** and it's an active process too).

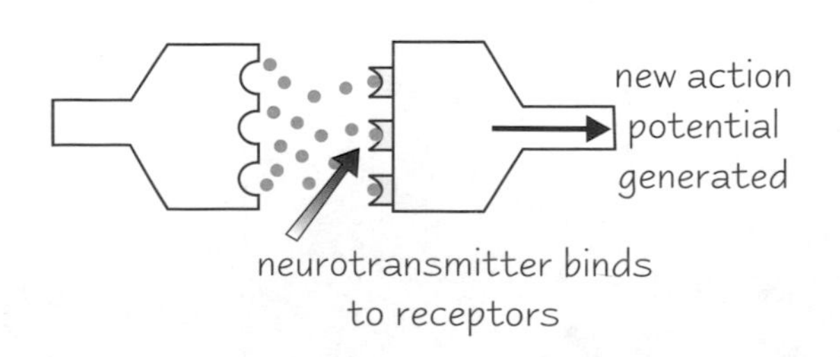

4) The neurotransmitter **diffuses** across the synaptic cleft and binds to **receptors** on the **postsynaptic membrane** of the other neurone.

5) This stimulates an **action potential** in the postsynaptic membrane by opening the **sodium channels** (see page 20).

6) An **enzyme** is sometimes used to **hydrolyse** the neurotransmitter, so the response doesn't keep on happening. The neurotransmitter may also be taken back up into the presynaptic bouton, ready to be used again.

- Because the receptors are only on the **postsynaptic** membranes, a signal can only pass across a synapse in **one direction** (it's **unidirectional**).
- The postsynaptic cell behaves as a **transducer**, just like **receptor cells**, because the **chemical** stimulus (neurotransmitter) is converted into an **electrical** one (action potential).
- There are lots of **mitochondria** in the bouton of a neurone. These provide the **ATP** to make more neurotransmitter, power exocytosis and pump calcium ions out of the bouton.

Synapses

There are *Different Kinds* of *Neurotransmitters*

There are many kinds of neurotransmitters. A lot of neurotransmitters are **excitatory** (as described above), but some are **inhibitory**. These make the membrane **hyperpolarised**, creating a resting potential value that's even more **negative** than the usual -65 mV. This makes it harder to **excite**.

Some kinds of neurones in particular parts of the body, only have receptors for certain **types** of neurotransmitter, e.g:

1) The neurotransmitter **acetylcholine** binds to **cholinergic receptors**. It's an **excitatory** neurotransmitter at many neurones and neuromuscular junctions.
2) **Noradrenaline** binds to **adrenergic receptors** and it's generally **excitatory**.

Some **drugs** and **poisons** affect the action of neurotransmitters:

- **Agonists** are chemicals that **mimic** the effects of neurotransmitters. An example is **nicotine**, which binds to **cholinergic** receptors and stimulates the **neuromuscular junction**, connecting motor neurones to muscle fibres.
- **Antagonists block** the effects of neurotransmitters. An example is **curare**, a chemical that binds to **cholinergic receptors**, but blocks – rather than mimics – the acetylcholine molecules. This **paralyses** the muscle.

Synapses are sites of *Interaction* between *Different Signals*

The main function of a synapse is to bring about the passage of signals between neurones. However, another important function of synapses is that signals can **interact** there:

1) **Inhibition** of a signal. A neurotransmitter can **inhibit** a postsynaptic membrane and make it harder to excite by **hyperpolarising** it (see above).

2) **Summation** of signals. This means that the **overall effect** of lots of different neurotransmitters from lots of different neurones on one postsynaptic membrane is the **sum** of all their **individual effects** (bearing in mind that some might be **excitatory** and some might be **inhibitory**).

Practice Questions

Q1 What is the main function of a synapse?

Q2 What is exocytosis?

Q3 Why are there a lot of mitochondria in the bouton (synaptic knob) of the neurone?

Q4 Give two examples of a neurotransmitter.

Q5 What effect do antagonists have on neurotransmitters? Give an example of an antagonist.

Q6 What is summation at a synapse?

Exam Questions

Q1 Describe the sequence of events leading from the arrival of an action potential at a bouton to the generation of a new action potential on a post-synaptic membrane. [8 marks]

Q2 Explain how the structure of the synapse ensures that signals can only pass through it in one direction. [4 marks]

Bouton — like button, but in a weird French accent...

Not the most exciting page in the book, but pleasantly dull I'd say. There's nothing too hard there. Some chemicals with annoyingly long names cross a gap so action potentials can move from neurone to neurone through the body.

The Central Nervous System

And now ladies and gentlemen, the most important bit of all — the central nervous system, the controller of almost everything you do. It's like a giant computer, only much more complicated...

The Central Nervous System has Two Parts

The central nervous system is made up of your **spinal cord** and your **brain**:

1) The **spinal cord** runs through the protective, bony vertebral column. The **spinal nerves** containing sensory and motor neurones are connected to it. Quick **reflex** reactions like sneezing can be processed by the spinal cord, without using the brain.
2) The **brain** is a massive bunch of **relay neurones** (see p. 20) connected to the spinal cord. The brain is inside in the protective bony **cranium** (skull), so the nerves connected to the brain are called **cranial nerves**.

Nervous Tissue contains White and Grey Matter

White matter is mainly made up of the **nerve fibres** of the neurones — the **myelin sheaths** (see page 21) glisten, so it looks white.

Grey matter is a concentration of **cell bodies**, which appears darker because of the absence of the sheaths. In the spinal cord, the grey matter is in the centre, with white matter around the outside, but in the brain, the grey matter forms a thin surface coating called the **cortex**.

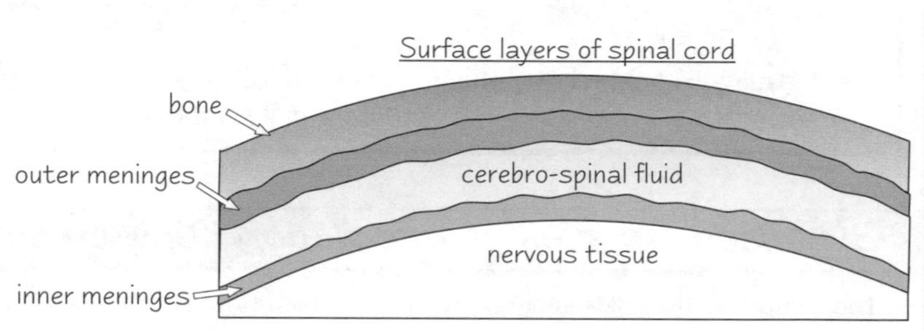

Running right through the spinal cord is a fluid-filled tube called a **sinus**, which carries **cerebro-spinal fluid**, produced by layers of connective tissue called **meninges**. The fluid absorbs mechanical shocks and supplies food and oxygen.

The Spinal Cord processes simple Reflexes

A **reflex** is an **involuntary stereotypical** response of part of an organism to an applied **stimulus**. This means that there's usually no **conscious control** over it, and it always produces the **same** kind of effect. It works because there are special patterns of neurones that make up **reflex arcs**. The simplest is a **monosynaptic reflex**, where the sensory neurone connects **directly** to the motor neurone so there's only **one synapse** within the central nervous system in the arc. An example is the **knee jerk reflex**. These reflex arcs often play a role in controlling **muscle tone** and **maintaining posture**. Action potentials don't pass to the **brain**, so no conscious thought is needed for them to happen.

A **polysynaptic reflex** has at least **two synapses** within the central nervous system, due to the presence of a **relay neurone**. Action potentials **can** pass to the brain, so some conscious thought might be involved. An example of such a reflex is the quick withdrawal response if you touch something hot.

Ventral roots and dorsal roots are just the parts of spinal nerves that join with the spinal cord.

These simple reflexes help you to **survive** because they mean you can respond very **quickly** to changes in the external environment. They **protect** the body from **harmful stimuli** and are there from birth, so you don't have to learn them.

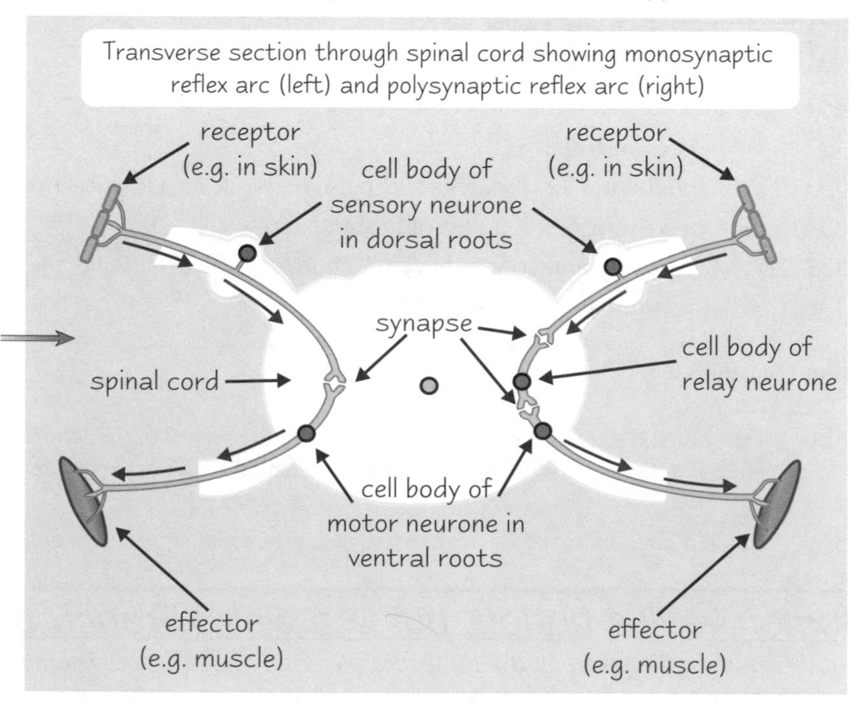

The Central Nervous System

*The **Lower Brain** controls **Unconscious Activities***

1) **Hypothalamus**
This is just beneath the middle part of the brain (above the medulla).
It controls important **conscious** activities, but it also **regulates** body temperature and water potential.
In some cases (e.g. water potential regulation), it communicates with the **endocrine system** (pages 18-19). The hypothalamus is the main part of the body where nervous and endocrine systems meet, because it is connected to the **pituitary gland** (see below).
Neurosecretory cells in the hypothalamus can carry action potentials and produce chemicals like ADH and oxytocin. Blood vessels carry these chemicals to the **pituitary gland** where they are stored and later released in response to action potentials from the hypothalamus.

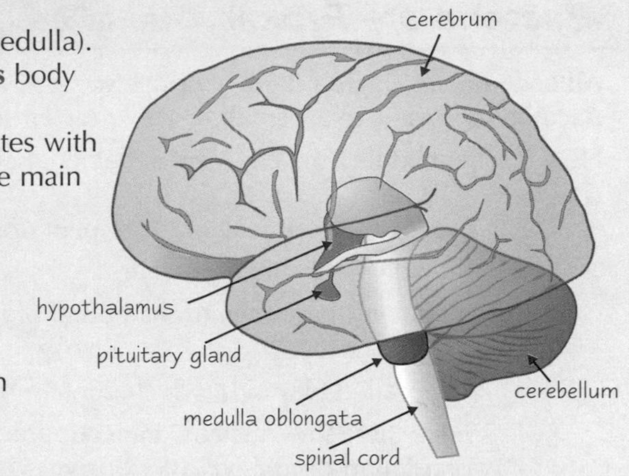

2) **Pituitary gland**
This controls the **endocrine system**, and releases a range of **hormones** (see pages 18-19 for more on this).
It responds to **neurosecretions** from the hypothalamus.

3) **Medulla oblongata**
This is at the base of the brain. It's the coordination centre of the **autonomic nervous system**.
It controls the unconscious activities of vital organs like the heart and lungs — it controls **breathing** movements, **heart rate** and also the action of **smooth muscle** in the gut.

*The **Cerebellum** and **Cerebrum** control **Complex Conscious Behaviour***

The **cerebellum** and the **cerebrum** each have a highly folded **cortex** (the outer bit). They coordinate sensory input with motor output. Action potentials arriving along sensory nerve fibres pass along lots of possible pathways via the relay neurones, so each **input** has lots of **different potential responses**. This is the basis of **memory** and learning from experience.

1) The **cerebellum** is important for balance and controlled muscle movements. It helps make movements coordinated. Over a period of time, **learning experiences** affecting the cerebrum influence the **neurone pathways** in the cerebellum. This means that the cerebellum becomes involved in controlling more **skilled coordinated sets of movements** that become routine (such as walking, maintaining posture or playing a piano).
2) The **cerebrum** is the largest part of the brain. It deals with the **voluntary activities** of the body. It's associated with advanced mental activity, like **emotion**, **memory** and **language**. It's divided into **two parts** (cerebral hemispheres). The right hemisphere deals with actions for the left side of the body, and vice versa.

Practice Questions

Q1 Describe the basic structure of the brain and spinal cord.

Q2 How many synapses does a polysynaptic reflex arc have?

Q3 What is the function of simple reflexes?

Q4 Where is the hypothalamus located in the brain and what is its function?

Exam Questions

Q1 Describe a monosynaptic reflex arc, naming one example. [4 marks]

Q2 Explain why the pituitary gland is connected to the hypothalamus of the brain. [4 marks]

I'm not sure that I have a brain...

It goes without saying that the brain is a pretty complicated organ. That doesn't mean that it has to be too stressful to learn about it for your exam. All you have to do is know what each of the main areas do, and understand that the brain processes information from receptors and sends impulses to effectors. Don't worry about exactly what's going on there.

Bacteria

This section covers Option A of Unit 4 of the syllabus. *A **microorganism** is basically any organism that can only be seen with a **microscope**. They're mostly the single-celled type of beastie — bacteria, fungi, protoctista and viruses.*

Bacteria are Prokaryotic Microorganisms

All bacteria are **single-celled**, although some kinds of bacteria can stick together to make **clusters** or **chains**. Bacteria are really, really **small** — they range from about 0.1 to 2μm. They can be classified by their cell structure.
Here are the **features** shared by **most** bacterial cells:

1) They have a **cell wall** made of **peptidoglycan**.
2) They divide by **binary fission**.
3) They **don't** have **membrane-bound organelles** (e.g. a nucleus).
4) Their **DNA** consists of a circular molecule with **vital** genes (the bacterial **'chromosome'**), and smaller rings called **plasmids**, which contain other, non-vital genes, e.g. for antibiotic-resistance.
5) Since there's no nucleus, **transcription** occurs in the **cytoplasm**. **Translation** occurs on **70S ribosomes** (smaller than the 80S ones in eukaryotes).
6) **Genetic recombination** can occur by **conjugation**, where two cells join up to exchange parts of their DNA. Genes can also be **recombined** by exchanging plasmids, by transfer in viruses called **bacteriophages**, or by taking up DNA from the **surroundings** (e.g. from dead bacteria).

And here are some **special** features found in **only some** kinds of bacteria:

1) **Bacterial flagella** may be present.
2) **Mesosomes** are **folds** of the cell membrane — their purpose isn't known for sure, but they may be associated with respiration.
3) **Bristle**-like projections called **pili** and **fimbriae** allow the bacteria to **stick** to surfaces. (Fimbriae are **shorter** than pili and there are **more** of them.)
4) There may be a layer of **polysaccharide** (called a **glycocalyx**) around the cell wall. It could be a loose **slime** layer or a more solid **capsule**. It could **defend** the bacteria from white blood cells or stop it **drying out**.
5) Bacteria can be **heterotrophic** (consume organic molecules), **photoautotrophic** (make food using light energy) or **chemoautotrophic** (make food using energy from oxidation reactions).

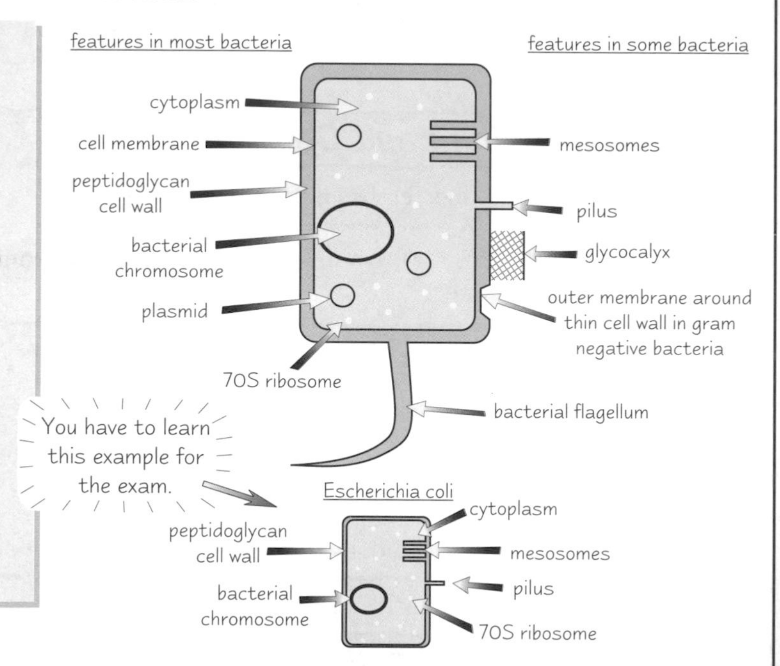

Bacteria are Classified on the Basis of Cell Wall Structure

Bacteria can be classified as either **Gram-positive** or **Gram-negative**:

1) **Gram-positive** bacteria have a **thick** cell wall. **Gram-negative** bacteria have a **thinner** cell wall coated with an outer **membrane**.
2) To test whether bacteria are Gram-negative or positive, they can be stained purple with **crystal violet** solution, then washed with Gram's **iodine** solution, then with **ethanol**, then counter-stained with the red stain **safranin**.
3) If, after this process, they end up **purple**, they're **Gram-positive**. If they end up **red**, they're **Gram-negative**.

Gram-positive wall

this stains purple with crystal violet (the red stain doesn't show over the purple)

thick cell wall

cytoplasm

phospholipid membrane

Gram-negative wall

the outer membrane of this (with the purple stain) is washed away by ethanol — the cell wall then stains red

thinner cell wall

cytoplasm

Bacteria

Some Types of **Bacteria** are **Pathogenic**

Some bacteria are **parasites** — they **invade** a living body and obtain **nourishment** from it. A **pathogen** is any microorganism that is a **parasite** of a plant or animal and causes **disease**. Whether bacteria are pathogenic depends largely on the structure of their **cell wall** and **capsule**. The cell wall and capsule enable the bacteria to:

- invade the body of the **host**
- **attach** to the cells of the host
- some species can even invade the individual **cells** of the host, causing disease

Tuberculosis is caused by the bacterium ***Mycobacterium tuberculosis***. It's contracted simply by breathing it in. The bacteria penetrate into **unactivated alveolar macrophages** (white blood cells found in the lungs), where they can stay **dormant** for years until the host's immune system becomes weakened, e.g. by old age. They can then replicate inside the macrophages and spread to **lymph nodes** and other tissues. Affected tissues are damaged and symptoms include weight loss and coughing.

Bacteria that are better equipped to invade a host and cause a disease there are said to have high **pathogenicity**. The **number** of bacteria needed to cause a certain disease is an indication of the **infectivity** of the bacteria. The **invasiveness** is the ability of bacteria to **spread** within the host.

Many Pathogenic Bacteria Produce **Harmful Toxins**

Most kinds of pathogenic bacteria produce chemicals as **by-products** of their metabolism. Chemicals that can cause harm inside the body of the host are called **toxins**.

Exotoxins are soluble chemicals that are **released** from the cells. They are produced by bacteria such as ***Staphylococcus*** and ***Vibrio cholerae***, which causes cholera.

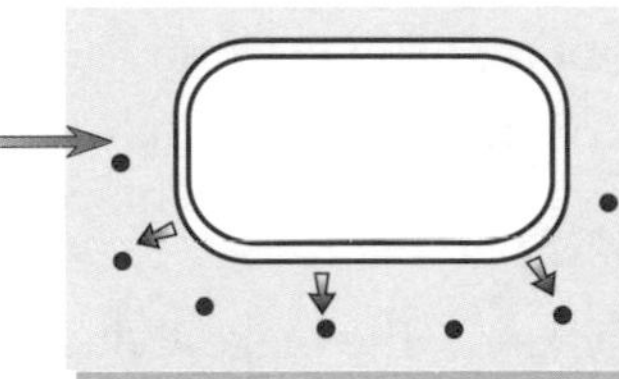

Exotoxins released from bacterium (e.g. Staphylococcus aureus) quickly cause food poisoning.

Endotoxins are **retained** inside the bacterial cells, but are released when the cell walls and membranes of these bacteria are damaged. ***Salmonella enteriditis*** produces endotoxins that prevent absorption of water in the large intestine, leading to **diarrhoea**.

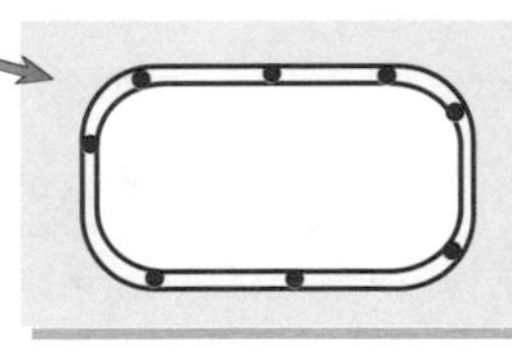

Endotoxins inside cell, released on death of bacterium (e.g. Salmonella enteritidis) cause food poisoning after a delay.

Usually **higher** concentrations of endotoxins than exotoxins are needed to produce the symptoms of disease.

Practice Questions

Q1 List the basic features of bacterial cells.

Q2 Describe the procedure used to distinguish between Gram-positive and Gram-negative bacteria.

Q3 Describe how *Mycobacterium* bacteria can invade the body to cause tuberculosis.

Q4 Distinguish between exotoxins and endotoxins, naming an example of bacteria that produce each type.

Exam Questions

Q1 Describe how the bacteria *Salmonella enteritidis* can bring about food poisoning. [2 marks]

Q2 a) Describe two possible functions of the layer of polysaccharide around a bacterial cell wall. [2 marks]
b) Draw a labelled diagram to show the layers around the cell wall of a Gram-negative bacterium. [3 marks]

Bacteria — Unhygenix's wife...

A lovely couple of pages on bacteria. A lot of this stuff shouldn't be too new. You'll know a lot about bacteria and their microorganism friends from all your past Biology studies. It makes it a bit easier for you, but it doesn't mean you can get by with what you know already. So go on — learn those nasty, long words. You know you want to.

Fungi and Viruses

Fungi exist in one of two forms, multicellular, e.g. moulds or unicellular, e.g. yeasts. Viruses are different from bacteria and fungi — they're acellular (not made of cells). They're not even living things. So don't ever, ever say they are.

Fungi are *Eukaryotic* Organisms

Some types of fungi are filamentous and **multicellular**, like moulds, and others are smaller and even **single-celled**, like yeasts.

Here are the general characteristics of **moulds**:

1) These fungi are made up of many thread-like structures called **hyphae**. Together, these hyphae form large **webs** called **mycelia**. An example of a mould is *Penicillium* (used to produce penicillin).
2) Cell walls are made of **chitin**. Most fungi have partition walls called **septa** along the length of hyphae.
3) **Vacuoles** are usually present in the cells of hyphae.
4) Cells are **diploid** and divide by **mitosis**. Genetic recombination occurs by **meiosis** to form **spores**, and **conjugation** (where DNA is passed between cells).

A fungal hypha

Yeasts are single-celled fungi, e.g. *Saccharomyces*. Unlike **moulds** and other types of fungi, they have cell walls made from complex **polymers of mannose** (a sugar) and **glucose**. Yeast cells divide by **budding**, involving a kind of asymmetrical mitosis.

A budding yeast cell

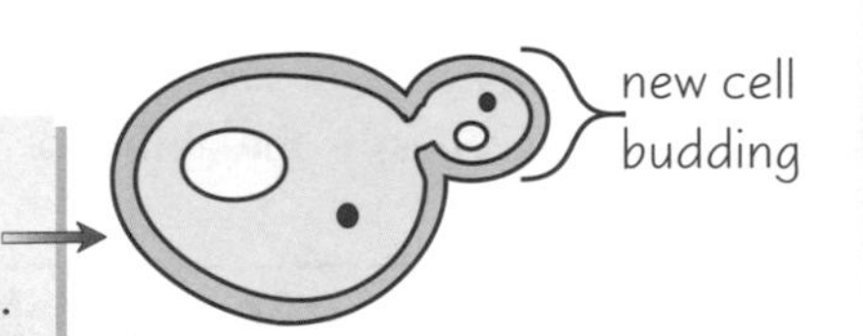

Viruses are Just Particles of *Nucleic Acid* and *Protein*

Viruses are microorganisms that are **not** made of cells, so can be described as **acellular**.

All viruses are **pathogens** (infectious organisms that usually cause disease), whereas only **some** types of bacteria, fungi and protoctists are pathogens. Some viruses only infect bacteria. These are called **bacteriophages**, or sometimes just **phages** for short.

Viruses are **tiny**, even compared to bacteria. The biggest virus is only 0.2 μm across and most are much smaller. This means that they can only be seen with an **electron microscope**.

Classification of viruses is based on virus **structure** and the type of **nucleic acid** they contain.

Their basic **features** are:

1) A core of **nucleic acid**, **either** DNA **or** RNA (not both), sometimes with some **protein**.
2) An outer coating of **protein** called a **capsid**, made up of protein units called **capsomeres**.
3) All viruses are **infectious** — they invade cells.
4) Some viruses have an extra layer, called an **envelope**, made of **membrane** from the cell membrane of a previous host cell.
5) They have **no** organelles, **no** cytoplasm and **no** plasma membrane.

Capsids and **envelopes** may:

1) **Protect** the virus from **chemicals** (e.g. enzymes) while outside a host.
2) Enable the virus to **bind** to host cell membranes, ready for **penetration**.
3) Assist in **penetration** of a host cell.

The structure of three different viruses:

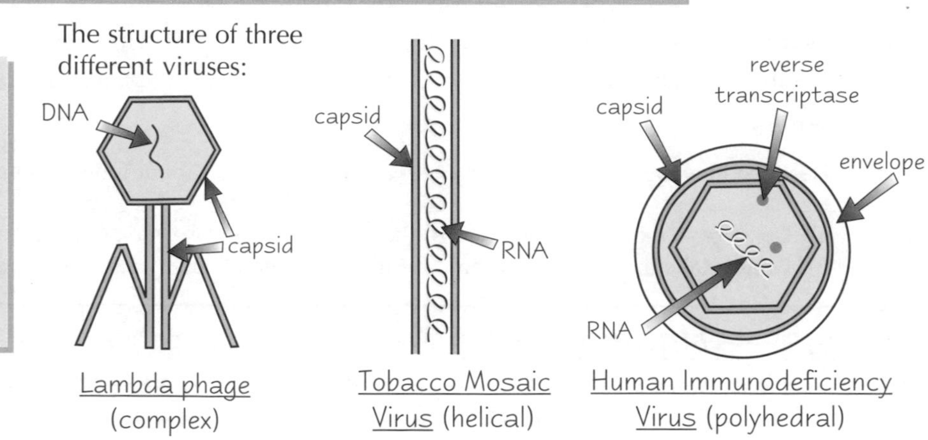

Lambda phage (complex) — Tobacco Mosaic Virus (helical) — Human Immunodeficiency Virus (polyhedral)

Fungi and Viruses

*Viruses **Replicate** Inside their **Host Cells***

Viruses can only **reproduce** and make **protein** inside a **host** cell. They don't have the equipment, such as **enzymes** and **ribosomes**, for doing this on their own, so they use those of the **host**. The **life cycle** of a virus is an '**infection cycle**':

1) The capsid or envelope of the virus particle **attaches** to a **receptor** molecule on the host cell membrane.
2) The virus **penetrates** the cell and the capsid **uncoats** to release the **nucleic acid** into the host cell's cytoplasm.
3) The nucleic acid **replicates** and is **transcribed** to make **mRNA**. **Translation** occurs on the host's **ribosomes** to make more capsid **proteins**.
4) New viruses are **assembled**. Then they are **released** from the cell by **lysis** (splitting) of the cell or by **budding** from the cell.

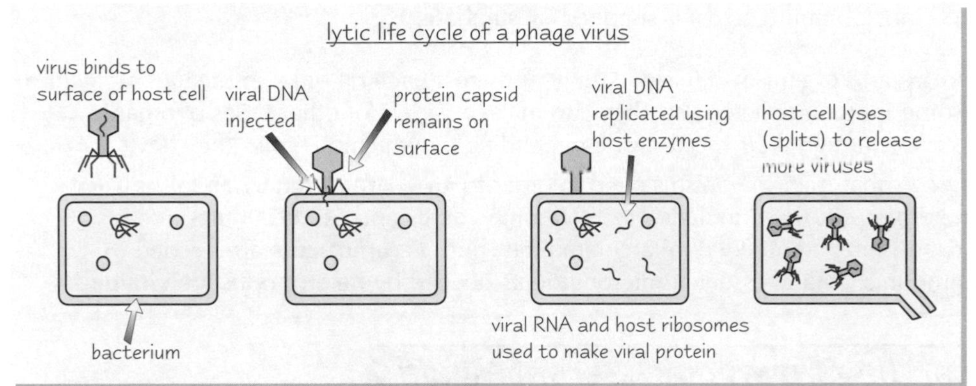

This diagram shows an infection cycle for phage viruses but it's the same process for all viruses.

HIV** is the Virus which leads to **AIDS

Human immunodeficiency virus (HIV) affects the white blood cells known as **helper T-cells** in the immune system. It eventually leads to **acquired immune deficiency syndrome (AIDS)**. AIDS is a condition where the immune response worsens and eventually fails. This makes the sufferer more **vulnerable** to other infections, such as pneumonia.

HIV is a **retrovirus**. This means it carries an enzyme called **reverse transcriptase**, which does this:

1) Inside the white blood cell, reverse transcriptase is used to make a **complementary strand** of DNA from the **viral RNA template**.
2) From this, **double-stranded DNA** is made, which is **inserted** into the human DNA.
3) The DNA stays in the white blood cell for a long while (the **latency period**).
4) Eventually, after a long period of time, the virus particles are **multiplied**. The DNA **codes** for the synthesis of **viral proteins** to make new viruses that burst out, killing the cell, and then **infect** and **destroy** other white blood cells.

Practice Questions

Q1 What are the differences in structure between yeasts and moulds?

Q2 List the basic features of a virus.

Q3 Describe the lytic life cycle of a phage virus.

Q4 HIV is a retrovirus. What does this mean?

Exam Questions

Q1 Explain what is meant by the term 'latency period' with reference to HIV infection. [2 marks]

Q2 Explain the functions of viral capsids and envelopes. [3 marks]

You can't avoid all viruses, so just learn to love 'em. Little darlings.

There are loads of different kinds of viruses, some are nastier than others. HIV is a very nasty one and we don't have a cure for it, only drugs to help slow it down. Viruses can cause all sorts of illnesses like cancer, flu, cold sores, stomach bugs and common colds (they're caused by rhinoviruses — nothing to do with real rhinos I'm afraid).

Cell Culture

Loads of kinds of microorganism can be grown in the lab in cultures. Which means you have to learn all about it.

Microorganisms have Special ***Chemical Requirements***

Autotrophic organisms (plants, algae and some kinds of bacteria) need a supply of simple inorganic molecules with which to build organic molecules. So, if they're being grown in a lab, there should be a ready supply of these things. Here are the **elements** they need and where they get them from:

1) **Carbon** and **oxygen** (in all organic molecules) are supplied by **carbon dioxide** or water.
2) **Hydrogen** (also in all organic molecules) is supplied by **water**.
3) **Nitrogen** (in amino acids and the bases of nucleotides) is supplied by **nitrate**.
4) **Phosphorus** (in nucleotides) is supplied by **phosphate**.
5) **Sulphur** (in just two kinds of amino acids) is supplied by **sulphate**.

Heterotrophic organisms (animals, fungi and most bacteria) take in organic molecules, and some inorganic materials. They can make a variety of other types of organic molecules through their metabolism, depending on which enzymes they have.

Chemicals that microorganisms need in order to grow are called **essential nutrients**. For heterotrophs, these include **essential amino acids** and **essential fatty acids**. **Macronutrients** are needed in large amounts, but **micronutrients** are needed in tiny amounts. Many essential micronutrients needed by heterotrophs are **vitamins**.

Microorganisms can be ***Grown*** *on* ***Different Media***

Microorganisms are cultured for a variety of **reasons**, including:

- **identifying** the species of microorganism in a medical sample — to **diagnose** the disease
- producing **useful** substances, such as **antibiotics**
- **food production**, such as making cheese and wine

In the lab, microorganisms are cultured **in vitro** ('in glass', like on an agar plate). The material where microorganisms are grown is known as the **medium**. It can be liquid, in the form of a **nutrient broth**, or solid, as in **agar jelly**.

Agar jelly is made by heating a solution of agar, a polysaccharide, which solidifies to jelly when cooled to 42°C. Special nutrients are added before the agar sets.

Selective media contain substances which selectively **prevent** the growth of certain microorganisms, whilst allowing others to grow. This means that specific types of microorganism can be **isolated** and **cultured**.

Indicator media contain a chemical **indicator** (e.g. pH indicator) that changes colour due to chemicals made by certain types of microorganisms. For example, **EMB agar** contains lactose, sucrose, eosin and methylene blue. Methylene blue inhibits Gram negative gut bacteria and only some types of Gram positive ones can use lactose or sucrose. Of these, *Escherichia coli* produce black colonies with eosin, whereas *Salmonella* produce pink ones.

Other ***Conditions*** *are Needed for* ***Optimum Growth***

1) The optimum **temperature** for growth varies between microorganisms, with some having extremes. **Psychrophiles** grow best at low temperatures, **thermophiles** grow best at high temperatures.
2) Some microorganisms are **aerobic**, but others are **anaerobic**. **Obligate aerobes** can only respire aerobically and so will die if oxygen is not present. **Obligate anaerobes**, however, die in the **presence** of oxygen. If necessary, oxygen concentration in a culture can be increased by **aeration**.
3) Most species have an optimum **pH** of between 5 and 9, but **acidophilic** forms have a low optimum pH. Usually fungi are more tolerant of low pH than bacteria.
4) **Disinfectants** inhibit growth of microorganisms and **antibiotics** kill them or inhibit their replication. (See the next page.)

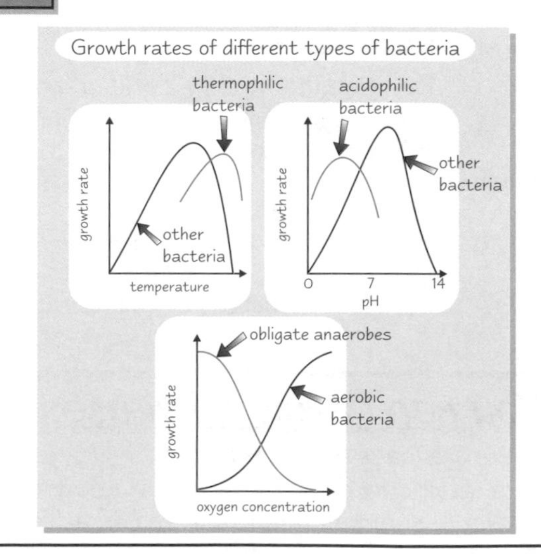

Cell Culture

Aseptic Technique is used to Culture Microorganisms

Aseptic technique is handling a microorganism culture in a way that minimises **contamination**. **Sterilisation** is the removal or killing of **all** microorganisms on an object, so that after sterilisation, there aren't any microorganisms to reproduce, even if the most favourable conditions occurred. The equipment and media used are sterilised by heating to 121°C for 15 minutes in an **autoclave**.

Other precautions taken to minimise **contamination** by other microorganisms include:

1) Lab workers pay close attention to **personal hygiene** (e.g. clean hands, hair net).
2) Windows and doors are kept **closed**.
3) All containers holding microorganisms have **lids** to minimise contamination by **'fall out'** of microorganisms from the **air**.
4) Working close to a **Bunsen flame** ensures that many of the microorganisms in 'fall out' are **killed** before they land on working surfaces.
5) Surfaces are swabbed with **disinfectant**.
6) Culture **containers** are made of smooth glass or stainless steel as **scratches** can harbour stray microorganisms.
7) The microorganisms can be transferred from medium to medium on **wire inoculating loops** that are first **sterilised** by heating them to red hot in a Bunsen flame. Alternatively, a sterilised **Pasteur pipette** can be used for transferring a known volume of culture (necessary for quantitative work), and the culture is then spread on agar using a sterilised **spreader** made from a bent glass rod. This gives a fairly even **lawn** of microorganisms.

Disinfectants and Antibiotics Affect Growth of Bacterial Lawns

The **effectiveness** of antibiotics and disinfectants at preventing bacterial growth can be discovered by adding them to bacterial lawns on sterile discs of filter paper.

The disinfectant or antibiotic **diffuses** out of the disc to create an **inhibition zone**, where bacteria can't grow. The **bigger** this is, the more **effective** the antibiotic or disinfectant.

Practice Questions

Q1 Name five elements which microorganisms need to build organic molecules. Where do they get them from?

Q2 Explain the meaning of the terms 'selective media' and 'indicator media'.

Q3 Explain the difference between obligate aerobes and obligate anaerobes.

Q4 What is meant by sterilisation of laboratory apparatus?

Q5 How can you test how good a disinfectant or antibiotic is?

Exam Questions

Q1 A student was provided with a monoculture of the mould *Aspergillus nidulans* on an agar plate in a Petri dish. She was asked to transfer a sample of the *A. nidulans* culture onto a separate sterile agar plate.

a) Describe the stages in the aseptic transfer of the sample from the monoculture to the sterile new plate. [5 marks]
b) Suggest a method that was used to sterilise the agar plate. [2 marks]

Q2 a) Explain why a culture of bacteria in a nutrient broth showed increased population growth when oxygen was bubbled through the medium. [3 marks]
b) Suggest how the results would be different for a bacterial culture of an obligate anaerobe. [2 marks]

Glasgow, City of Culture 1998 — Great conditions for bacterial growth...

This is the sort of stuff that's really boring... but really important. I know you want to get on with learning about nasty diseases and other gruesome stuff, but you've got to take the time for cell cultures. If it wasn't for cell cultures, we wouldn't be able to diagnose lots of diseases, or make antibiotics to cure them, or, erm, make cheese. So there.

Growth of Cultures

It's a lot easier to measure the population growth rate of a bacterial culture, than of, say, lions. Or whales. Or people.

Bacterial Cultures Show Typical **Population Growth Curves**

Following the initial introduction of bacteria to the medium, the following **phases** of **population growth** are shown:

1) The **lag phase** occurs when there is a very **small** early **increase** in population density. **Reproduction rate** is **low** because it takes time for enzymes to be made for DNA replication and to use any new food.
2) The **exponential phase** occurs during the most **favourable** conditions. Here there is a **doubling** in population size per unit time (one cell divides to produce two cells, each of which can divide further in the same amount of time). There is sufficient **food** to support the growth, and **competition** for food is at a minimum.
3) The **stationary phase** is approached as **death rate** increases and becomes **equal** to the reproductive rate. This happens because **food** gets used up and poisonous **waste products** build up.
4) **Decline phase** occurs where death rate is **greater** than reproductive rate, due to the further depletion of food and accumulation of excreted waste.

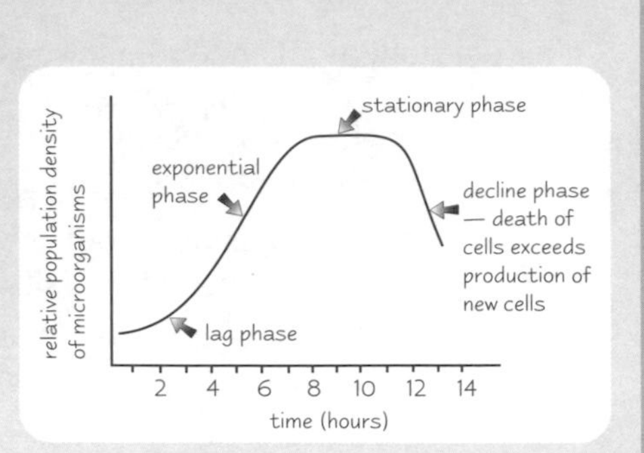

Depending on culture conditions, many bacteria produce substances called **secondary metabolites** towards the end of the exponential phase and into the stationary phase. These are not essential for growth and reproduction, but some, such as antibiotics, can help bacteria **survive** the stressful conditions at this point in the population growth curve.

Growth Rate can be Expressed by Calculating a **Growth Rate Constant**

Any **exponential increases** in a population size (or any other factor) can be **quantified** by working out the **$\log_{10}$ values** of the population sizes. If we plot the $\log_{10}$ values of the population size of an exponentially growing population on a graph, we end up with a **straight line**.

This kind of conversion then lets us calculate something called the **growth rate constant**. It's equal to the number of **generations** that the population goes through in a particular length of time. It's calculated by using the following **formula**.

$$\text{growth rate constant} = \frac{\log_{10} N_1 - \log_{10} N_0}{t \times \log_{10} 2}$$

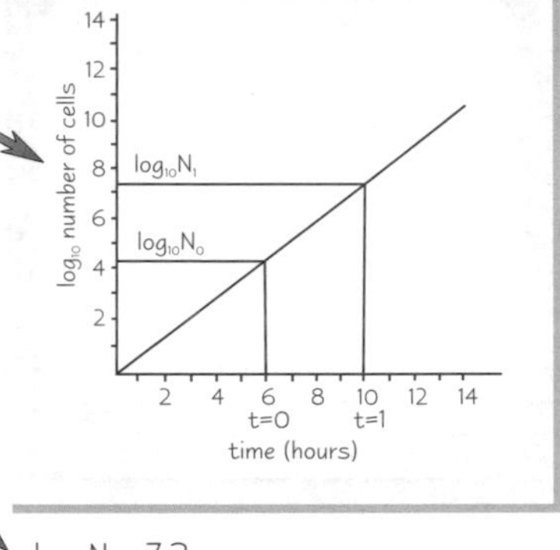

Look at the straight line graph shown above.
Here's what the **formula** means:

- **$\log_{10} N_1$** is the log of the number of cells at a point in time (**t = 1**)
- **$\log_{10} N_0$** is the log of the number of cells at an earlier point in time (**t = 0**)
- **t** is the **time** taken between these points

$\log_{10} N_1 = 7.3$
$\log_{10} N_0 = 4.3$
$t = 10 - 6 = 4$ hours

therefore the growth constant rate is:

$$\frac{7.3 - 4.3}{4 \times 0.301} = \frac{3}{1.204} = 2.49$$

Because this is a log graph, you can simply **read them** off from the axes — you **don't** need to calculate them. Similarly, $\log_{10}$ of 2 = 0.301. (You won't even need to work this out; any question on this will give you the $\log_{10}$ value of 2.) The calculation value of the growth rate constant in this time interval is shown under the graph.

Bacteria with **Two Food Sources** often Show **Two Exponential Phases**

If bacteria are grown on media with **two** different **carbon sources**, they will often use one before starting on the other. This is called **diauxic growth**.

E. coli grown on a mixture of **glucose** and **lactose** will use up the glucose first. This is because the presence of glucose **represses** the gene that codes for the production of the enzyme **lactase**. The lactase is finally made during the second lag phase when all the glucose is used up.

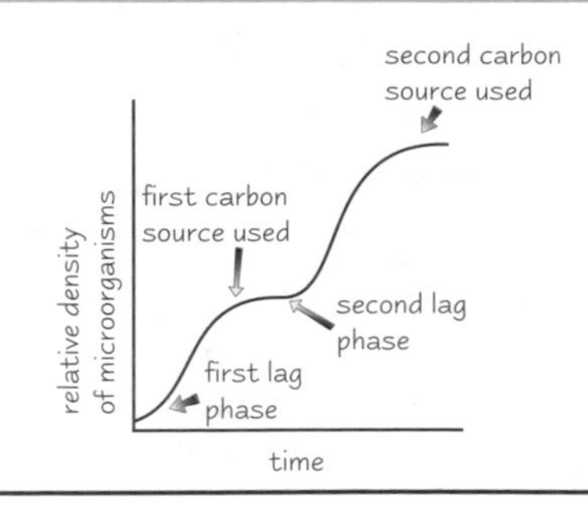

Growth of Cultures

There are Different Ways to Find out **Population Sizes** of **Bacterial Cultures**

Some techniques are designed to determine the actual **population density**.
This involves working out the number or mass of **cells** in a particular volume of medium.

1) Direct **cell counts** are possible by taking a tiny sample of the medium and examining it under a **microscope**. You can then count the number of cells you see in a fixed volume of the medium using a **haemocytometer** (a grid of microscopic chambers, each with a known area and depth). It's best to do a number of counts and find the **average**, for it to be **accurate**. This number can then be **multiplied** to give the number in the sample.

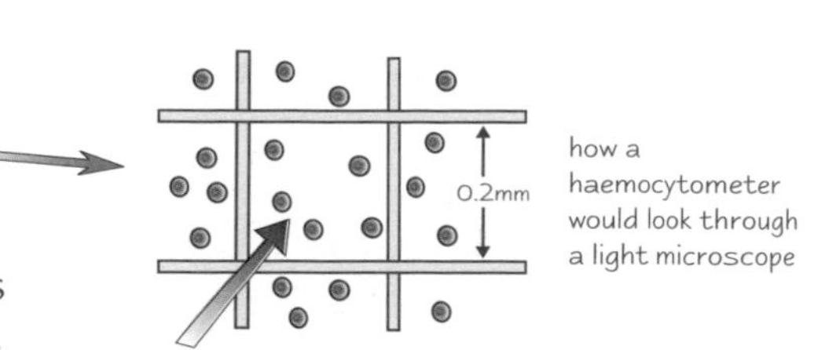

2) The **density** of cells can be measured with a **turbidometer** (colorimeter) — the more **turbid** (cloudy) the medium is, the more light it absorbs, so the less **light** is transmitted through the turbidometer. The turbidity is then **compared** to the turbidity of a **known** number of bacteria.
3) It's possible to measure **fungal** cultures by simply **weighing** them, but this isn't usually sensitive enough for bacteria.

The **Dilution Plating Method** tells us the Number of **Viable Cells**

All the methods above tell us the total **quantity** of cells in a sample of culture. The **dilution plating method**, however, can tell us the number of **viable** (living) **cells** in a medium — those that can divide. Here's how it's done:

1) A known volume of the culture is **diluted** by a known amount. It's then **further diluted**, again by a known amount each time.
2) A known volume of the final dilution is spread over an **agar plate** and **incubated**. Each single viable cell found in this final dilution will **reproduce** to produce a visible **colony**.
3) The number of colonies is **counted**. By knowing the **dilution factors** used each time, it is possible to work out the **total** number of cells in the first sample.

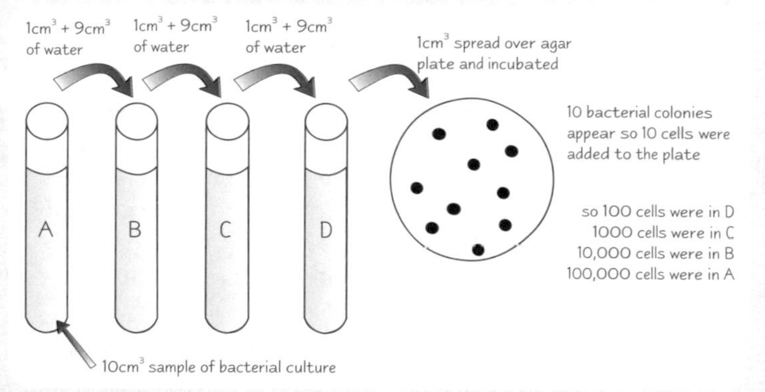

Practice Questions

Q1 Which factors result in lag, exponential, stationary and decline phases in a bacterial population growth curve?
Q2 What is meant by the term 'secondary metabolite'? Give an example of a secondary metabolite.
Q3 Describe the basis for the turbidity technique in establishing the density of a bacterial culture.
Q4 How would you count the number of viable cells in a sample?

Exam Questions

Q1 A bacterial culture was grown in a laboratory fermenter for 10 hours.
a) Calculate the growth rate constant where $\log_{10} N_1 = 8.6$, $\log_{10} N_0 = 3.4$ and $\log_{10}$ of $2 = 0.301$. [3 marks]
b) Outline the dilution plating method that was used to establish the density of bacterial cells at the two points in time. [8 marks]

Q2 a) What is meant by diauxic growth? [2 marks]
b) Draw a graph for a bacterial culture that illustrates diauxic growth, labelling the phases. [3 marks]

Log graphs? OK, that's it — I want out...

Now, don't be scared off by that nasty-looking formula. It's really not that bad — all you have to do is read your figures off the graph and then plug them into the formula. Or if you're lucky, you might be given the figures to just plug in. Anyway, you've dealt with much worse in your GCSEs so you can definitely handle it. And then the rest of the page is dead easy.

Industrial Growth of Microorganisms

The large-scale culturing of microorganisms means we can produce lots of important substances... like yoghurt. Great.

Many **Industries** *Grow Microorganisms on a* **Large Scale**

All sorts of microorganisms can be grown to provide useful **products**. Many of these products are important in the **food industry**, such as alcohol in wine-making or lactic acid in yoghurt manufacture. Others are grown for **medical** applications, such as for antibiotic production.

These substances are produced by the metabolic reactions of the microorganisms.

Large cultures of microorganisms are grown in vessels called **bioreactors**. The culture methods used are designed to **increase** the growth rate of microorganisms through creating **optimum conditions** (see page 30).

There are two main **culture methods** that are used:

Batch culture occurs in a **fixed volume** of medium. **Oxygen** is usually added during the growth of the microorganism and **waste gases** removed. Growth occurs up to the **stationary** phase until food becomes depleted and excretory products accumulate. Eventually **new cultures** must be established to start the process over again in a different batch. It is easier to **control** conditions in a batch, and **isn't costly** to start again if contamination occurs.

In **continuous culture**, fresh, sterile medium is added to the culture at a constant rate. Used-up medium and dead cells are removed at a constant rate. Cells are kept in the **exponential** growth phase and the culture lasts for far **longer** than with the batch method. This method is more **productive** than batch culture and **smaller** vessels can be used. However, if **contamination** occurs, the whole lot will be lost, which is **very costly**.

Conditions *in a* **Bioreactor** *are Controlled to* **Optimise Microorganism Growth**

A typical **bioreactor** needs these conditions to successfully grow microorganisms:

1) **Aseptic entry of material** — if the bioreactor is contaminated the whole batch will have to be discarded.
2) **Culture innoculant** — the medium needs to be innoculated with the starter culture (this just means the culture is added to the medium). The culture needs to have the properties you want, e.g. high production of penicillin.
3) **Suitable medium** — the medium has to contain all the essential nutrients needed for the microorganism that is being cultivated.
4) **Aeration** — if the organisms are **aerobic** they will need a plentiful supply of oxygen for **respiration**. **Sterile air** is delivered through an **aerator**.
5) **Temperature** — the correct temperature is needed for optimum enzyme activity. When organisms respire they produce a large amount of heat. Bioreactors have **cooling jackets** to control temperature. Cool water is circulated through the **jacket**, which reduces the risk of **overheating**, especially in larger bioreactors. The heat released by fermentation is transferred to the cool water in the jacket by conduction and is carried away as the water flows out.
6) **pH** — a constant pH is needed for optimum enzyme activity. Bioreactors have a **pH controller**. The controller contains a **pH probe** to detect changes in pH (e.g. caused by the acids produced by fermentation). It automatically delivers a **neutralising** quantity of **acid** or **base** (an alkaline solution) when required.
7) **Agitation** — the culture needs to be **stirred** to bring fresh medium into contact with the microorganisms. In bioreactors a motor drives the rotation of an **impeller**, which **stirs** the culture.
8) **Product recovery** — there needs to be an effective way of removing the product. This is usually through a **tap**.

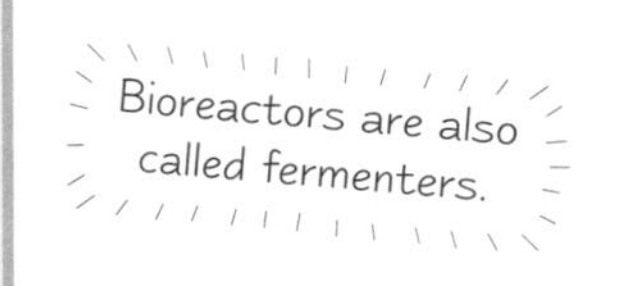

Industrial Growth of Microorganisms

Bioreactors need to be kept **Sterile**

Large-scale fermenters are difficult to keep **sterile** and heating and cooling can be slower, making it more difficult to **control** conditions.

The whole bioreactor is usually made of smooth stainless steel. This can be sterilised before use by pumping **steam** through the apparatus.

After this, the product is **purified** and sometimes chemically **modified**.
This series of treatments after harvesting is called **downstream processing**.

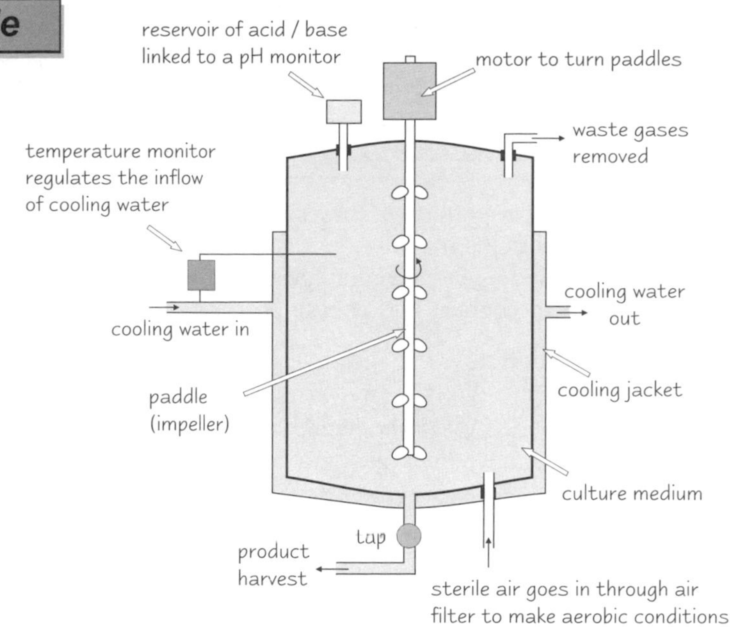

Penicillin is Produced by a Specific Type of **Mould**

Penicillin is an **antibiotic** made by a mould called *Penicillium chrysogenum*. It's grown in **bioreactors** by **batch culture**.

Here's how penicillin is made:

1) Organic **carbon** and **nitrogen** sources are added to the medium, as well as **oxygen** for aerobic respiration.
2) **Downstream processing** of the penicillin begins by **filtering** the mould mycelium from the medium.
3) The penicillin is **extracted** in an organic solvent.
4) It is **concentrated** and **crystallised**.
5) It may be chemically **modified** so that its effects can overcome penicillin-resistant strains of bacteria.

Mycoprotein is also produced using a type of mould called *Fusarium graminareum*.
Mycoprotein is compacted mycelium, you'll probably know it as **Quorn ™**. It's produced in a **bioreactor** by **continuous culture** (see page 37 for more on Quorn production).

Practice Questions

Q1 Describe how each of the following components of an industrial bioreactor will optimise the growth rate of the cultured microorganism: water cooling jacket, pH controller, stirrer.

Q2 How are bioreactors kept sterile?

Q3 What is meant by the term 'downstream processing'? Describe the downstream processing of penicillin.

Q4 Name two kinds of fungi that are grown in industrial fermenters. What useful products do they synthesise?

Exam Question

Q1 a) Distinguish between the terms batch culture and continuous culture. [2 marks]
b) Give one advantage each for batch culture and continuous culture. [2 marks]
c) Explain why the antibiotic production of a *Penicillium* bioreactor would decrease if the flow of sterile air into the reaction vessel were interrupted. [3 marks]

Quorn — truly a gift from the gods for all the veggies of the world...

Sausages... mmm... bacon... mmm... chicken fillets... mmm... spare ribs... mmm... ham slices... Yep, there's absolutely nowt they can't create with Quorn these days. And it tastes good too. Anyway, all this stuff is fairly straightforward again. Just remember the bioreactor bits and that it's all about making sure as many microorganisms are grown as possible.

Biotechnology in Food Production

Mmmm... yoghurt, bread, beer, Quorn. Mmmm... beer. Made by fermenting yeast? Mmmm... yeast.

Some microorganisms produce useful by-products when they respire, e.g. lactic acid, ethanol and carbon dioxide. We often use these microorganisms to help produce foods that contain these products.

Bacterial Fermentations are used in Yoghurt Production

When microorganisms are cultured, they often make a substance known as a **fermentation product**. This fermentation product is made under anaerobic conditions, when the respiratory substrate is **not** completely **oxidised**. Two common fermentation products are **lactic acid** and **ethanal**. These are important in **yoghurt** and **cheese** production.

Here's how **yoghurt** is made:

1) Milk is screened for **pathogenic bacteria** and **antibiotics**. (If antibiotics have been given to cows, they may be present in the milk.)
2) It is **heated** to around 90°C for 30 minutes to **kill** bacteria, **denature** milk proteins and **reduce** oxygen levels.
3) It is **cooled** to 45°C. Lower temperature and oxygen levels are needed for the **starter culture**.
4) **Starter culture** is added, including *Lactobacillus bulgaricus* and *Streptococcus thermophilus*. These are **incubated** at around 32°C to 40°C. *Lactobacillus* **hydrolyses** proteins to peptides, (which are used by the *Streptococcus*), and *Streptococcus* makes **methanoic acid** (for growth of *Lactobacillus*). *Lactobacillus* produces **lactic acid**, so the pH drops and a sour flavour develops. Both produce **ethanal**, which provides more **flavour**.
5) Sterilised **flavourings** (e.g. fruit) are added.
6) The yoghurt is **stored** at 2°C to reduce activity of the bacteria.

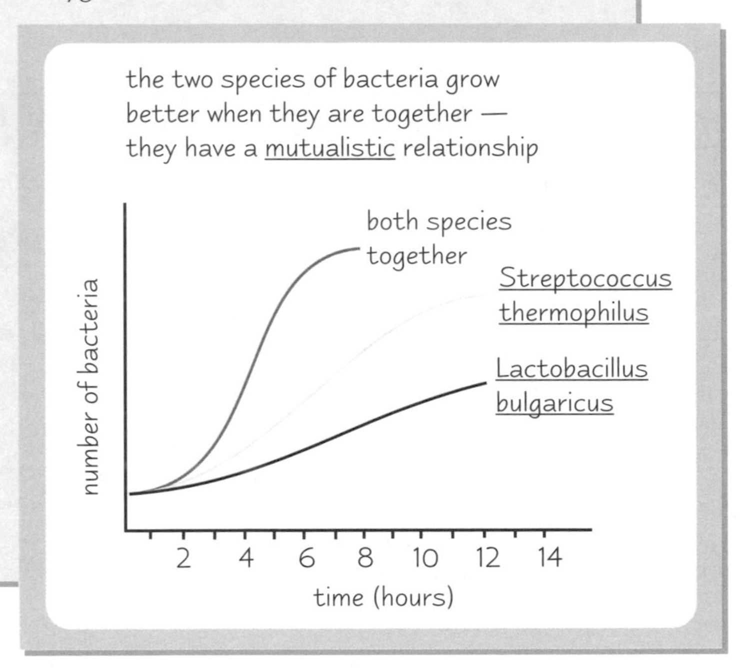

Yeast Fermentations are used in Brewing...

The **yeast** *Saccharomyces* is used in brewing **beer**. Its fermentation product is **ethanol**.

1) Germinating barley seeds (the **malt**) contain both **starch** and the **amylase** enzyme needed for the hydrolysis of starch into soluble sugars. When this malt is **mashed**, it produces a sweet **liquor**.
2) The liquor is **fermented** by yeasts, which switch to anaerobic respiration as oxygen is used up — so converting sugars to ethanol and carbon dioxide.
   - **Top fermentation** involves *Saccharomyces cerevisiae*, which produces a **froth** on the surface of the fermenting vessel. This is how **ales** are made.
   - **Bottom fermentation** uses *S. carlsbergensis*, producing **less froth** so yeast sinks to the bottom. This takes longer at lower temperatures and makes lighter **lagers**.
3) Yeast is **removed**, the surplus being used in the food industry.
4) The beer is **conditioned** before sale.

Biotechnology in Food Production

...and also in Bread-Making

The yeast *S. cerevisiae* is also used in dough to produce light (leavened) **bread**. Enzymes in wheat flour **hydrolyse** starch, mainly to maltose and some glucose. The yeast converts these sugars to carbon dioxide and some ethanol. It is the **carbon dioxide** that makes the bread **rise**. As the carbon dioxide expands, it gets trapped in the dough, making it lighter.

Mycoprotein is a Food Produced from Another Type of Mould

Mycoprotein is compacted mycelium (see p. 28) with a similar consistency to **meat**.
You'll know it by the name of **Quorn** — a vegetarian source of **protein**.
It's obtained from the growth of a **mould** called *Fusarium graminareum* by **continuous culture**:

1) A **continuous culture** of the mould is grown in a **bioreactor** at around 30°C and at a slightly acidic pH. **Carbon** is provided as glucose syrup, and **nitrogen** as gaseous ammonia. Vitamins, salts and oxygen are also given.
2) Large quantities of **protein** are made, but so are large amounts of **RNA**, which could be harmful. This RNA is **reduced** by adding **ribonuclease**, an enzyme that hydrolyses RNA to nucleotides.
3) The hyphae are **filtered** and **compacted** into a 'cake', which is sold as **Quorn**.

Practice Questions

Q1 Name a product of bacterial fermentation and one of yeast fermentation that are used in food production.
Q2 Name both bacteria used in yoghurt production.
Q3 Describe the role of yeast in brewing.
Q4 Explain how yeast causes bread to rise.
Q5 Describe how mycoprotein is produced.

Exam Questions

Q1 Brewing begins with the mashing of malt (germinating barley seeds) to produce sweet liquor.
a) Describe the processes that occur in order to produce the sweetness. [3 marks]
b) A yeast culture is used to act on the sweet liquor to produce ethanol. This requires anaerobic conditions. Explain how these conditions arise in the fermentation vessel. [2 marks]
c) Suggest how the quality of the brew would be different if the concentration of oxygen in the culture was higher. Explain your answer. [2 marks]

Q2 Yoghurt production uses a starter culture of two kinds of bacteria:
Lactobacillus bulgaricus and *Streptococcus thermophilus*.
a) Describe the role of each in affecting the flavour of the yoghurt. [4 marks]
b) Explain how these kinds of bacteria are mutually supportive. [4 marks]
c) Flavourings, such as fruit, are often added at the end of yoghurt production. Explain why these must be sterilised first. [2 marks]

Mmm... yoghurt... a combination of bacteria and gone-off milk... great...

Here's a useful tip for you: If you ever decide to make bread and it goes a bit wrong and isn't cooked properly, don't feed the nasty uncooked mess to the ducks. The reason why (and it's a good one) — ducks can't burp, so when the yeast carries on making carbon dioxide inside them, they'll eventually explode. Erm, don't try this out just to see if it's true either...

Antibiotics

These pages are all about antibiotics — what they do and where they come from. There's also a bit about antibiotic resistance, which is quite scary — the evolution of super-bacteria which aren't killed off by antibiotics. Eeek.

Antibiotics Kill or Prevent the Growth of Microorganisms

Antibiotics are **chemicals** produced naturally by a range of different microorganisms (especially fungi and bacteria). They either **kill** or inhibit the **growth** of other types of microorganisms.

- **Microbicidal** antibiotics **kill** microorganisms.
- **Microbistatic** antibiotics inhibit **growth** of microorganisms.

Antibiotics are **secondary products** of the **metabolic reactions** of a microorganism, and are released from their cells. Although the microorganism uses up **energy** producing the antibiotic, it is an **advantage** to do so because antibiotics prevent the growth of other kinds of microorganisms. This reduces the numbers of **competitors** for food.

Antibiotics are often very **specific** in their mode of action. They usually work by acting as **inhibitors** of enzymes and other types of proteins. Some antibiotics are so **damaging** to the affected cells that the cells are **killed**.

Antibiotics Inhibit Enzymes Involved in Crucial Metabolic Reactions

The reason why many antibiotics are so **effective** at reducing growth or killing microorganisms is because they **inhibit** the metabolic reactions that are **crucial** to the growth and life of the cell.

1) Some inhibit enzymes that are needed to make the chemical **bonds** in bacterial **cell walls**. This stops the cells from **growing** properly. An example is **penicillin**.
2) Some inhibit **protein synthesis** by binding to bacterial **ribosomes**. Examples are **tetracycline** and **streptomycin**.
3) Some inhibit **nucleic acid synthesis** by binding to bacterial **DNA** or **RNA polymerase**. An example is **rifampicin**.

These antibiotics can be used to treat **diseases** caused by bacteria. They are able to do this because they bind to molecules that are **only** found in **bacterial** cells, rather than the host cells.

For example, bacteria have **70S ribosomes**, whereas the eukaryotic host cells have different shaped **80S ribosomes**. This means that the antibiotic will work to **kill** the **bacteria**, but will leave the host cells alone.

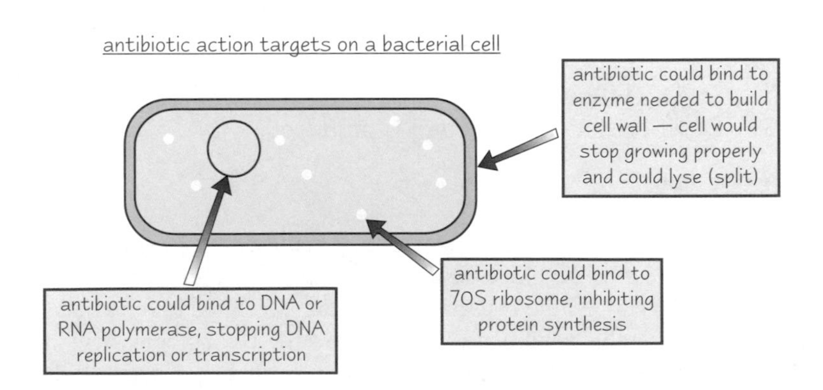

Antibiotics Vary in their Range of Effects

Some antibiotics have a **broader range** of effects than others:

- Antibiotics that target **specific** reactions in **specific** microorganisms are called **narrow spectrum antibiotics**.
- Other antibiotics inhibit enzymes that occur in a **wide range** of microorganisms, and so can be used to treat a **wide range** of different disease-causing pathogens. These are called **broad spectrum antibiotics**.

Antibiotics

Antibiotics can be Made in Bioreactors

1) The bacteria and fungi that produce antibiotics are **cultured** in industrial **bioreactors** in order to produce **large quantities** of particular antibiotics for use in **medicine**.
2) The conditions in the bioreactor need to be **controlled** in order to optimise the growth rate of the microorganism.
3) Because antibiotics are secondary metabolic products, they tend to be produced in the **later** stages of the **exponential** growth phase.

(See pages 34-35 for more).

Mutated Genes Cause Antibiotic Resistance

Microorganisms that are resistant to antibiotics have randomly **mutated genes** causing a change in their chemical makeup. The antibiotic is prevented from **binding** to the enzyme, but the reason why depends on the type of **mutation**.

For example, bacteria that are resistant to **penicillin** have a gene allowing them to produce an enzyme called **penicillinase**. This **breaks down** the penicillin molecules before they can inhibit the enzyme that builds the bacterial cell wall.

Antibiotic Resistance Spreads Quickly Through Bacterial Populations

Bacteria reproduce very quickly, and every time the cells divide, the DNA replicates. **Natural selection** causes the incidence of antibiotic resistance to rise quickly. This is because exposure of a population to an antibiotic **kills** those bacteria **without** the antibiotic resistance gene. Then the **resistant** ones, which obviously survive, can reproduce without **competition**.

Genes for resistance can also spread because of the exchange of **plasmids**. Plasmids are small **rings of DNA** in bacterial cells and many contain antibiotic-resistance genes. Plasmids are exchanged when two bacteria join together in a process called **conjugation**.

Overuse of antibiotics has resulted in **antibiotic resistance** being a big problem in medicine. There are currently certain strains of disease-causing bacteria (e.g. the *Mycobacterium* that causes tuberculosis) that have **evolved** resistance to most common antibiotics. This makes the treatment of these diseases increasingly difficult.

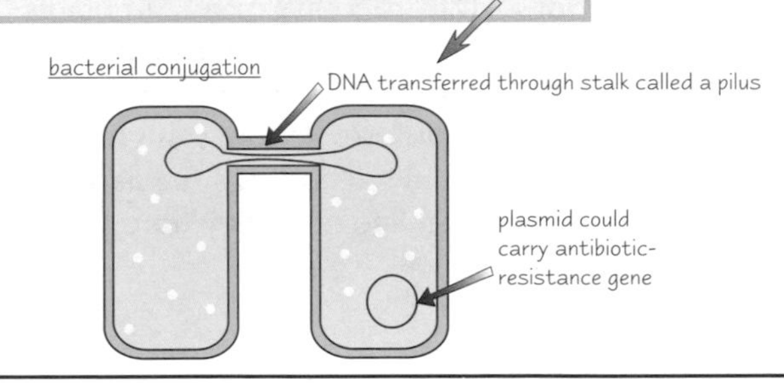

Practice Questions

Q1 Distinguish between the effects of the antibiotics penicillin, tetracycline and rifampicin.

Q2 Explain why antibiotics that target protein synthesis will only affect prokaryotic (bacterial) cells, rather than the eukaryotic host cells.

Q3 What is the difference between broad spectrum and narrow spectrum antibiotics?

Q4 Antibiotics are described as secondary metabolites. What does this mean?

Exam Questions

Q1 Explain how resistance for penicillin can spread through a population of bacteria. [4 marks]

Q2 What is the difference between a microbistatic antibiotic and a microbicidal one? [2 marks]

My Auntie Biotic's really well cultured...

(oh shut up now with the stupid lines on stupid things... they're not at all funny)

Hooray! You've finished this section! So what was your favourite bit? I quite liked page 36 because now I feel I know how to brew beer. I wouldn't advise having a go at home though, partly because I shouldn't condone an underage interest in alcohol, but also because the dog always knocks over your big brewing vessels and the resulting mess just stinks.

Balanced Diet

This section covers Option B of Unit 4 of the syllabus.
Food is made up of a mixture of carbohydrates, lipids, proteins, vitamins and minerals. They're all really important for being healthy.

Food is a **Complex Mixture** of **Organic** and **Inorganic** Compounds

Organic food compounds belong to **four main groups**:

1) **Carbohydrates** include **sugars** and **polysaccharides** (e.g. starch).
2) **Lipids** include fats and oils, and the **fatty acids** they're made of.
3) **Proteins** and the **amino acids** they're made up of.
4) **Vitamins**.

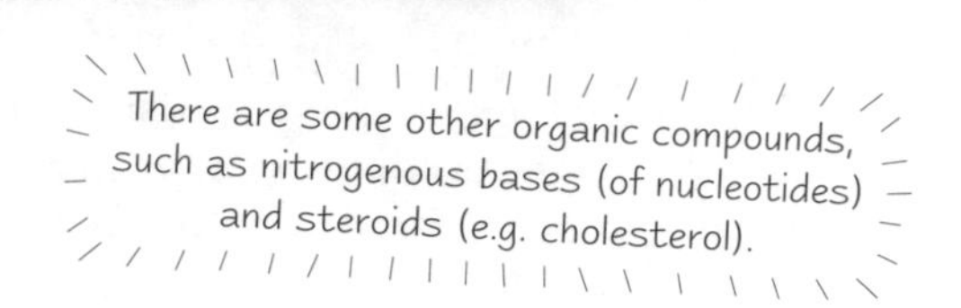

And there are two types of inorganic compound:

5) **Mineral salts** contain useful **mineral ions**.
6) **Water**.

A **balanced diet** has all these different compounds in the right proportions for good health. Of course, these proportions are different for each person — they change because of things like age, gender and lifestyle. Some compounds can be made by the cells in the body, but you've got to get the others from your diet — these are called **essential nutrients** and they include vitamins, minerals and some types of amino acids and fatty acids.

Carbohydrates are our **Main Source** of **Energy**

Simple **sugars** are **monosaccharides** or **disaccharides** — these are soluble, sweet-tasting carbohydrates. **Glucose** (a monosaccharide) is used by all cells in respiration to release energy. Disaccharides such as **sucrose** (your bog standard sugar at home) make up most of the sugar intake in our diet. They've got to be hydrolysed into monosaccharides, like glucose, in the small intestine before they can be absorbed and used.

Long-chain carbohydrates (loads of monosaccharides stuck together) are **polysaccharides**. They're found mainly in plants — **starch** and **cellulose** are the main polysaccharides in plant matter. We hydrolyse starch in our gut into glucose and absorb that. But we can't break down cellulose. It goes straight through us, cleaning our insides a bit, keeping us regular ... nice. Sometimes it's called roughage or dietary fibre.

Sugar	Dietary Source
Glucose	grapes, other sweet fruits, onions
Fructose	sweet fruits, honey
Sucrose	principal plant transport sugar, found in sap
Maltose	germinating seeds

Lipids provide **Energy** too, but are also Needed for **Other Specific Tasks**

Lipids are **fats** and **oils** — they're compounds of glycerol and fatty acids. Mass-for-mass they contain more energy than carbohydrates. A **saturated** fatty acid has no carbon-carbon double bonds, **mono-unsaturated** has one, and polyunsaturated has two or more. **Polyunsaturated** fatty acids are the more healthy ones.

H–C(H)(H)–C(H)(H)–C(H)(H)–C(H)(H)–COOH

saturated fatty acid — has no C=C double bonds

H–C(H)(H)–C(H)=C(H)–C(H)(H)–COOH

single C=C double bond in monounsaturated fatty acid

H–C(H)=C(H)–C(H)=C(H)–COOH

more than one C=C double bond in polyunsaturated fatty acid

Fats are solid at room temperature, oils are liquid.

Phospholipids are the main molecules in **cell membranes** and they also help to carry lipids in the blood stream. (Lipids are absorbed first into the lymphatic system before moving into the blood — some are carried with proteins as lipoproteins.) **Steroids** are also fatty substances. The most famous is **cholesterol**, needed in cell membranes and to make certain hormones and bile salts — it's a real celebrity.

Balanced Diet

Many Different things Change your rate of Energy Release

The **metabolic rate** is the speed of all the reactions inside your cells. It varies a lot depending on what the body is doing. The **basal metabolic rate (BMR)** is a better comparative measurement — it's the rate of energy release when the body is at rest. It's worked out by finding the rate of **thermogenesis** (heat loss from the body, using a human calorimeter) or from the rate of **oxygen consumption**, and is usually expressed per kg of body mass.
BMR varies with:

1) **Gender**: BMR is higher in males than females. Females have a higher proportion of **adipose** (fatty) **tissue**, which respires at a lower rate. If the adipose tissue effect is removed and BMR is just based on **lean body mass**, there is no difference in BMR between the sexes. Interesting...
2) **Age**: BMR falls as a person gets older
3) **Body mass**: BMR increases with increasing body mass (due to more respiring tissue)
4) **Disease**: fever makes BMR increase, as can an overactive thyroid gland
5) **Pregnancy**: BMR is greater during pregnancy and lactation
6) **Food intake**: digestion increases BMR

Exercise involves lots of muscle contraction, which uses up lots of energy. This means that you need to make more energy from your food. More intense exercise demands more energy, pretty obviously. So regular exercise increases your BMR.

Type of activity	Amount of energy spent / MJ per day
Sedentary work	4.0
Moderate work	5.5
Heavy work	7.5

Essential Nutrients need to come from your Diet

Proteins that you eat are hydrolysed in the gut into amino acids. These are absorbed into the blood and carried off to the cells to make proteins you need. Some kinds of amino acids can be made from other types in the cells; others — the **essential amino acids** — must be provided in the diet.

Vitamins are organic compounds that are needed in very small quantities to maintain health, often by acting as **cofactors** (molecules that help enzymes work properly). **Mineral ions** may act as cofactors too. An important one is **sodium** (from sodium chloride), needed to maintain the osmotic properties of blood and tissue fluid, and necessary to generate nerve impulses.

Essential Nutrient	Food Source
Vitamin A (retinol)	Fish liver oil, dairy products, liver, green vegetables
Vitamin C (ascorbic acid)	Citrus fruits, green vegetables, potatoes
Vitamin D (calciferol)	Fish liver oil, butter, eggs
Calcium	milk, cheese, bread
Iron	meat, liver
Iodine	Seafoods

Practice Questions

Q1 What is a balanced diet?

Q2 What are carbohydrates and lipids used for in the body?

Q3 What's the difference between metabolic rate and basal metabolic rate?

Q4 Why are some amino acids described as essential nutrients?

Exam Questions

Q1 a) What is the basal metabolic rate (BMR)? [1 mark]
b) Explain why rate of oxygen consumption and heat loss can be used as indicators of BMR. [2 marks]
c) Explain why BMR increases with increasing body mass. [2 marks]

Q2 Suggest one way in which the balanced diet of a two year old child would be different from that of a forty year old. Give a reason for the difference. [2 marks]

Q3 Explain why a diet rich in protein could lead to increased growth of working muscle, as well as increased urea production. [8 marks]

A balanced diet — a kilo of chocolate for every kilo of chips...

During the making of these pages, I've eaten a bar of Dairy Milk, two chocolate digestives, a slice of toast with jam and a banana. I've also had three cups of tea and a Diet Coke. Meanwhile, I've been sitting in one chair, and not moving much except to eat. So, anyway, I can't lecture you on healthy eating and exercise, but I can tell you that you need to know it.

Under-Nutrition

Vitamins, protein, calcium, iron and iodine are all essential to stay healthy. If you don't get enough of them you could be in trouble... bleeding gums and bow legs here we come...

A Lack of Protein can cause Starvation

A lack of protein in the diet is called **protein-calorie malnutrition**. It's common in the developing world. The diets of people in poorer countries may be particularly low in protein if protein-rich foods are expensive. Also, children need a greater proportion of protein than adults (so they can grow), so they may suffer more.

Conditions caused by lack of protein:

1) **Marasmus** is malnutrition caused by a diet that is low in protein and low in carbohydrate. Sufferers are very very underweight.
2) **Kwashiorkor** is malnutrition caused by low protein only. A common symptom is **oedema**. This is a swelling of tissues due to blood protein levels being very low. If blood protein levels are low tissue fluid will not re-enter the capillaries at the venous end (as it normally does) because the blood does not exert a big enough osmotic effect. The fluid accumulates, causing swelling. It can also reduce the production of digestive enzymes, so absorption of food from the gut is lowered.

This kwashiorkor sufferer has oedema of the abdomen

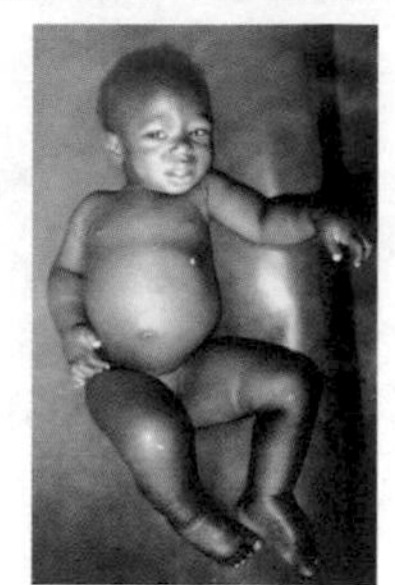

Malnutrition also occurs due to psychological disorders such as **anorexia nervosa** and **bulimia nervosa**. Anorexia nervosa results from an obsession with weight loss, leading to self-starvation. Bulimia nervosa involves bouts of binge eating, followed by self-induced vomiting. Both can be fatal.

Vitamin A is needed for healthy Lining Tissue and Eyes

Vitamin A (retinol) is a fat-soluble vitamin found in dairy products and fish liver oils. The body can make vitamin A from a group of compounds called the **carotenoids**, found in orange and yellow coloured fruit and vegetables (e.g. carrots).
Vitamin A is needed to keep lining tissue (epithelium) healthy. It's also used to make the visual pigment called **rhodopsin**, found in the rod cells of the retina. Vitamin A deficiency leads to night blindness and in severe cases **xerophthalmia** — where the cornea hardens and scars, causing total blindness. The skin also becomes hard and flaky, making it prone to infection. Nasty.

Vitamin C is needed to maintain healthy Connective Tissue

Vitamin C (ascorbic acid) is a water-soluble vitamin especially found in citrus fruits, green vegetables and potatoes. Vitamin C is needed for the enzyme **hydroxylase** to work properly.

Hydroxylase is essential in maintaining the **collagen** fibres in **connective tissue**. It converts an amino acid called **proline** into another one called **hydroxyproline**. **Hydrogen bonds** hold the polypeptide chains of the collagen fibres together. These bonds can only form from one hydroxyproline to another hydroxyproline on a different fibre.

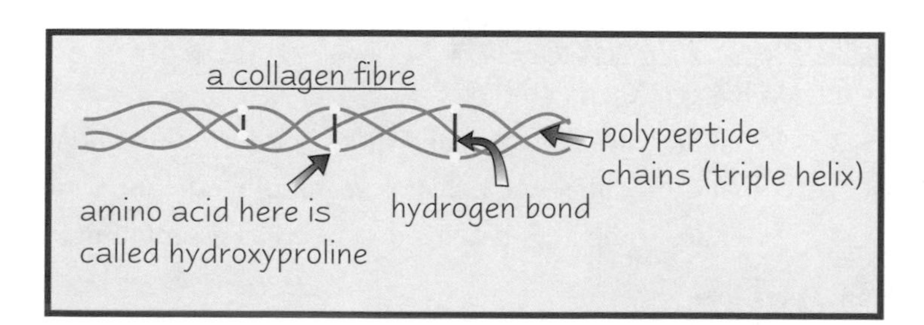

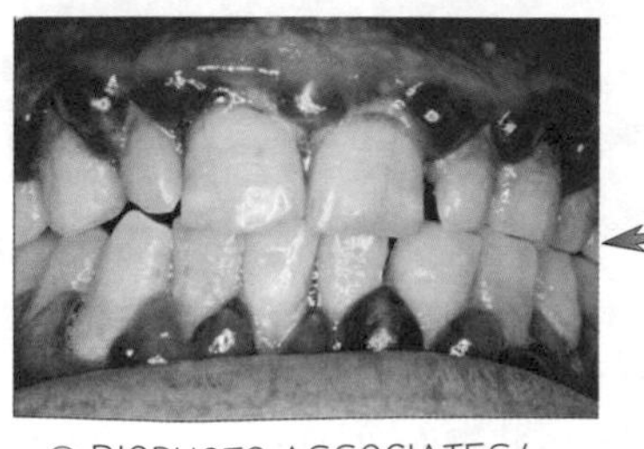

© BIOPHOTO ASSOCIATES/ SCIENCE PHOTO LIBRARY

Deficiency of vitamin C means that hydroxylase won't work properly and hydroxyproline is not formed. This means the hydrogen bonds can't form and the **connective tissue weakens**. This can lead to a condition called **scurvy** which causes bleeding gums, bleeding under the skin, bleeding around the joints and poor wound healing.

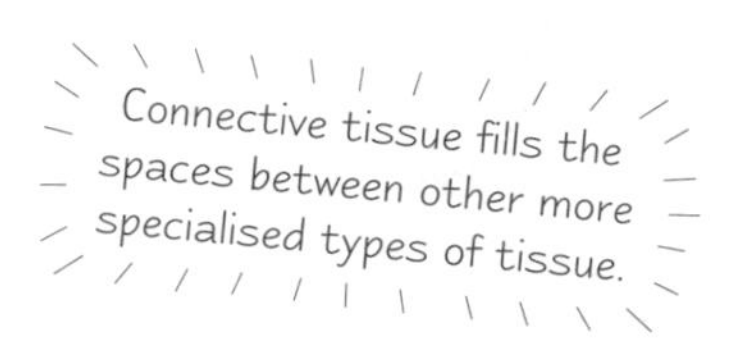

Under-Nutrition

Calcium is needed for healthy development of Bones and Teeth

Calcium is found in **dairy** products, **bread** and **flour**. It's absorbed from the small intestine through a **protein carrier**. This carrier can only be made if **vitamin D** is present in the diet. In fact, people can suffer from the effects of calcium deficiency because of lack of vitamin D, even though there might be plenty of calcium in the diet.
Calcium deficiency in the diet leads to **rickets** in children and fracture-prone bones in adults (**osteomalacia**).

Calcium is also needed for **nerve** and **muscle** function — it plays an important role at synapses (see pages 22-23) and in muscle contraction. This means that a deficiency in calcium (or vitamin D) can cause muscular spasms too.

Some chemicals (like **oxalate**) make calcium insoluble, which reduces its absorption. Spinach, for example, contains oxalate which reacts with calcium in this way. So while vitamin D is an **enhancer** of calcium absorption, oxalate is an **inhibitor**.

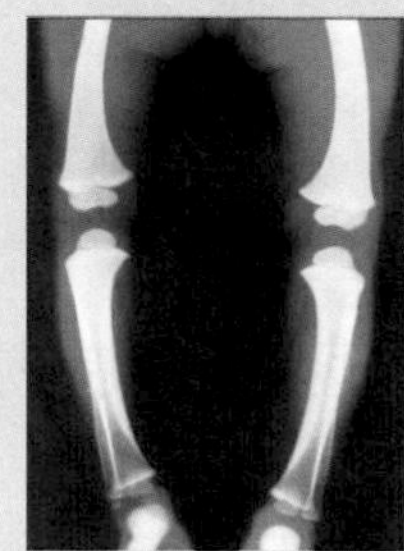

Bow legs are a common symptom of rickets.

Iron is needed to make Haemoglobin

Iron is found in both meat and plants. It's needed for **haemoglobin** in red blood cells, so a deficiency of iron causes a form of **anaemia**, due to insufficient haemoglobin in the blood. Sufferers of anaemia are visibly paler and feel tired due to the fact that less oxygen is being carried to respiring cells.

Iron **absorption** is affected by various factors:

Iron from meat is found in an organic molecule called **haem**, which normally exists bound to a protein, e.g. haemoglobin. This haem-iron is more **effectively absorbed** than the non-haem iron of plants, so you tend to get less iron from a **vegetarian diet**. But, absorption of haem-iron is still affected by various **enhancers** or **inhibitors** (similar to what happens with calcium). Absorption also depends on the amount of iron **already stored** — if this amount suddenly goes down, e.g. during menstruation, the body absorbs a lot more.

Practice Questions

Q1 What are the possible effects of vitamin A deficiency?

Q2 What is the role of vitamin C in the manufacture of collagen?

Q3 Explain why deficiency of vitamin D may lead to problems with bone development.

Q4 Name two essential minerals that are affected by absorption enhancers and inhibitors.

Exam Questions

Q1 a) Explain why people with vitamin C deficiency have poor wound healing. [4 marks]
b) Suggest two good sources of vitamin C. [2 marks]

Q2 Explain why a low protein diet could lead to oedema (fluid build up in the tissues). [4 marks]

Q3 Explain why a strict vegetarian diet could lead to anaemia. [6 marks]

So carrots make you see in the dark...

That's not totally true, but you can make vitamin A from a chemical in carrots, and that helps your eyes. There are loads of vitamins and minerals here to learn. You need to know about them, or you might accidentally get nasty problems from not eating the right things — like the student who got scurvy from eating nothing but porridge. Yuck.

Over-Nutrition

Loads of interesting stuff here on why it's bad to stuff your face with food.

Find your **Body Mass Index** to see if you're overweight or obese

The body mass index (BMI) is the official way of saying whether someone is overweight or obese.

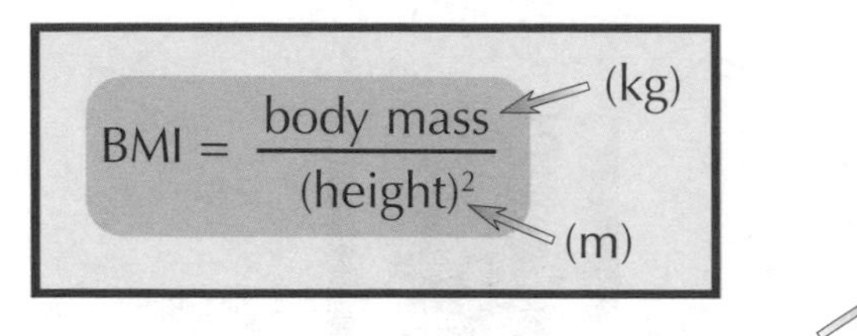

The table shows how BMI is used to classify people's weight.

Body Mass Index	Weight Description
below 18.5	underweight
18.5 - 24.9	normal
25 - 29.9	overweight
30 - 40	moderately obese
above 40	severely obese

Obesity reduces your **Life Expectancy**

The risk of early death is higher if you're obese. You're more likely to get **coronary heart disease**, **high blood pressure** and you're at an increased risk of **stroke** and some **cancers** (e.g. breast, cervical and colon). There may also be a higher risk of certain other conditions, such as **osteoarthritis**, **gallstones** and **diabetes type 2**. Not good at all.

Coronary heart disease is often caused by **atherosclerosis** (get ready for lots of silly words) ...

Atherosclerosis

1) **Fatty deposits** called **atheromas** (also called atheromatous plaques) form beneath the artery endothelium lining, **narrowing** the vessel. This reduces the blood flow as there is increased resistance.
2) The fatty material mainly comes from cholesterol in the plasma, carried by **low-density lipoproteins (LDL)**.
3) If atherosclerosis occurs in the coronary arteries it can cause **angina pectoris** (chest pains).
4) If the membrane over the plaque ruptures (breaks open) then blood cells stick to the exposed material. This forms a blood clot. If this clot blocks the arteries supplying the heart it causes **myocardial infarction** (heart attack) and if it blocks arteries in the brain it causes **stroke**. Uh-oh.

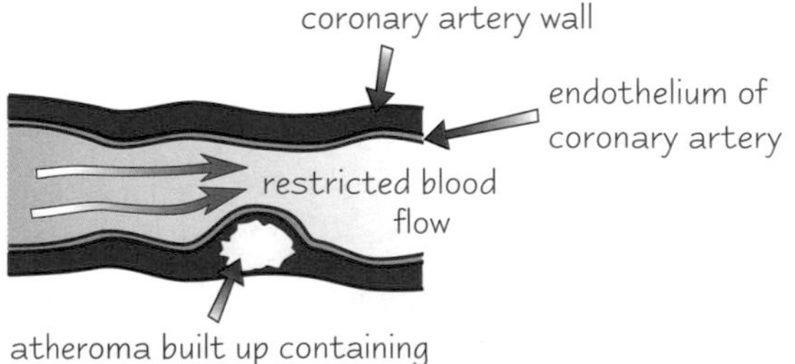

Saturated fat in the diet can **increase** the risk of atherosclerosis (which is bad), whereas polyunsaturates, e.g. vegetable oil (and soluble fibre, e.g. from porridge) have been shown to **reduce** plasma LDL and cholesterol levels (which is good).

The effects of excess salt on health are less clear, but it may cause high blood pressure.

Lack of **Dietary Fibre** can cause disease of your **Colon**

Dietary fibre mainly comes from **non-starch polysaccharides (NSPs)** which are found in the cell walls of plant matter (e.g. in apples and carrots). **Insoluble fibre** is largely cellulose, **soluble fibre** includes pectins and gums (carbohydrates found in cell walls of fruits). Fibre reduces the amount of time that food is inside the intestine (transit time) so there's less time for **harmful substances** in the food to affect the gut wall. If you don't have enough NSPs in your diet, you're more likely to get **constipation**, **irritable bowel syndrome** or even **cancer** of the large intestine. So who's for a big bowl of bran then?

Over-Nutrition

Over-nutrition can cause **Diabetes Mellitus** in adulthood

Diabetes mellitus is a condition associated with abnormally high levels of glucose in the blood.
There are two types:

Type 1 is caused by a **lack of insulin being secreted**. This is often diagnosed in childhood and must be treated with **insulin injections**.

Type 2 is caused by a **lack of insulin receptors** on liver cells, so they can't absorb glucose properly. This develops in middle-age and is often associated with **obesity**. Type 2 must be controlled by **diet**, rather than with insulin injections — sufferers produce insulin properly, but their liver cells can't respond to it.

Obesity is treated by **Exercising More** and **Eating Less** — pretty much

Actually, it's not quite as simple as this.

Exercise

The type of exercise is important — fairly **vigorous exercise** is needed to reduce body fat.

Dudley fitted some vigorous exercise into his lunch hour

Diet

Weight should be lost **gradually** by reducing **energy-dense** foods (like carbohydrates and lipids).

It's not a good idea to reduce **nutrient-dense** foods (such as those containing protein and vitamins) as it could lead to **deficiencies**.

Severe obesity may be treated using surgery, e.g. stapling of the stomach so less food can be eaten.

Practice Questions

Q1 Describe how body mass index is calculated, and how this value might give an indication of obesity.

Q2 What do these words mean?
a) atheroma b) angina pectoris c) myocardial infarction

Q3 What's the difference between type 1 and type 2 diabetes mellitus?

Q4 What are the main ways of tackling obesity?

Exam Questions

Q1 a) Explain how a diet high in saturated fat and cholesterol can lead to an increase in heart attack risk. [8 marks]
b) Name one factor that will lower blood cholesterol levels. [1 mark]

Q2 Fibre in the diet is said to decrease the 'transit time' through the intestine.
Explain what this means and why it is good for the health of the intestine. [3 marks]

Put your hands up and slowly step away from the cheeseburger...

Even if you're not overweight it doesn't mean you're free to eat whatever you want — if your diet consists solely of chips and pizza, you won't be doing your heart much good. There's some pretty weird names on these two pages... atherosclerosis, diabetes mellitus, myocardial infarction — make sure you know how to spell them and, ideally, what they mean too.

Food Additives

Two pages all about stuff we put in our food to try to make it tastier — but if you're having sprouts or liver for your dinner they aren't going to help you... nothing can.

Food Additives are in Loads of Foods

Food additives are chemicals that are added to processed foods to preserve them or to make them more palatable (i.e. improve flavour, texture or appearance). Food additives that are controlled by the European Union are given **E numbers**:

Additives		E Numbers
tartrazine (yellow)	colourings	E102
sunset yellow		E110
sulphites	preservatives	E221 to E227
sodium nitrite		E250
l-ascorbic acid	anti-oxidants	E300
tocopherols		E306
emulsifiers and stabilisers		E400 to E483

Some food manufacturers don't add additives to their products. This is because the **long-term effect** of eating some of them isn't known. Also, some people are sensitive to them and get rashes or upset stomachs if they eat them.

There are Different kinds of Sweeteners

The most obvious types of substances that taste sweet are **sugars** — monosaccharides like glucose and disaccharides like **sucrose**. But other things are sweet, and some of these are used as food additives.

Relative Sweetness

Sweetness is graded relative to that of **sucrose**, which is given a **relative sweetness** value of 1.0. Glucose is slightly less sweet (0.74), whereas fructose (sugar found in fruit) is much sweeter (1.73). The sweetness value is worked out by finding the lowest concentration at which the sweet taste can still be detected. This concentration will be greater for glucose than for fructose. Simple.

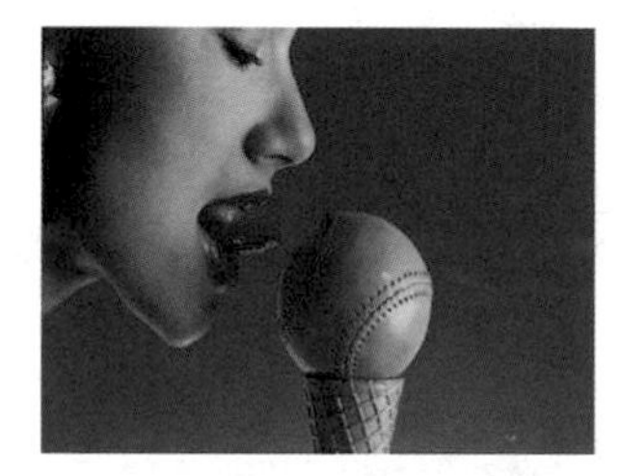

Jenny tested the relative sweetness of a baseball

Artificial sweetners can have very high relative sweetness values. An example is **aspartame**, with a relative sweetness value of 200. They're often used as low-calorie alternatives to sugars. Tiny amounts of aspartame can make food taste really sweet — it's often in those little sweetener things people put in their tea and coffee.
You need much more sugar to make something sweet, and that means more calories.

Many people have Lactose-Intolerance

If you don't have enough of the enzyme **lactase** in your intestine, you won't be able to hydrolyse lactose in milk properly — that's **lactose-intolerance**. Undigested lactose causes intestinal complaints (e.g. wind) because bacteria ferment it. Milk can be artificially treated with purified **lactase** in order to make it suitable for lactose-intolerant people. Lactose-intolerance is found in 10-15% of Northern Europeans, 50% of Mediterraneans, 95% of Asians and 80-95% of people of African descent.

Food Additives

Enzymes are used in *Food Processing*

During food processing, special enzymes can be used to change the carbohydrates (sugars), allowing manufacturers to **change the sweetness** of the food. The enzymes are produced by **microorganisms** grown in fermenters and are often **immobilised**. Immobilising them means that can be **reused** and are more **stable**, which makes the whole process cheaper.

Examples of such enzymes are **glucose isomerase** and **amyloglucosidase**...

Glucose isomerase converts **glucose** into sweeter **fructose**. Fructose is an **isomer** of glucose — it has the same molecular formula but a different structural formula (shape). This means that less sugar (and so fewer calories) needs to be used to give the same degree of relative sweetness.

glucose molecule

fructose molecule

Amyloglucosidase hydrolyses **starch** into **glucose** for use in sweet drinks. Sometimes this glucose is then treated with isomerase to make fructose. Great.

Practice Questions

Q1 Explain how relative sweetness of a substance is defined.

Q2 Compare the relative sweetness of sucrose, glucose, fructose and aspartame.

Q3 What does amyloglucosidase do?

Q4 Describe how fructose is made from glucose syrup using immobilised enzyme.

Exam Questions

Q1 a) Explain why lactose in the diet causes indigestion in people with lactose intolerance. [3 marks]

b) How can dairy foods be made safer for lactose-intolerant people? [2 marks]

Q2 a) The relative sweetness values for glucose, fructose and sucrose are 0.74, 1.73 and 1.0 respectively. Use this information to explain why fructose is considered to be a healthier alternative to sucrose. [3 marks]

b) High fructose corn syrup (HFCS) is used to sweeten many kinds of processed food. It is produced by immobilised glucose isomerase, which has an optimum activity at around 60 °C. Outline the advantages of using immobilised glucose isomerase of this temperature optimum. [4 marks]

Eee by gum, look at all the E numbers in that...

So there you go — some sugars are sweeter than others. You learn something new every page, don't you? Mmmm... sugar. But it's not all good — some people can't eat things with the sugar lactose in cos it makes them fart. Then there's those E numbers — they might taste yummy, but they make some people ill. And those blue Smarties, well, that's another story.

Other Additives

Here's some more pages on stuff that's put into food. There are some scary chemical names here but it's really all about making food look nice, taste nice and last for longer — you still need to learn the names though.

Processed Food often contains additional *Colourings*

Colourings are added to processed food to change the colour and make it look more appealing. Many colourings are substances that occur naturally, such as **carotene**, but others are man-made dyes, such as **sunset yellow** and **tartrazine**.

Recent medical findings have suggested that there is a link between eating tartrazine and incidence of hyperactivity in children. Most manufacturers don't use this additive anymore.

Antioxidants are added to *Preserve* food

Antioxidants stop food going off. Oxidation processes cause food to brown and spoiling of drinks. Food goes 'off' because it reacts with oxygen in the air. This causes lipids to become oxidised and long-chain fatty acids to turn into shorter chain fatty acids. These compounds make food taste bad. Many antioxidants are naturally occurring vitamins. Examples of antioxidants added to processed food include **L-ascorbic acid** (related to vitamin C) and **tocopherols** (vitamin E).

browning of apples due to oxidation

Flavourings and *Flavour Enhancers* are also added during Food Processing

Flavourings are added to processed food to improve taste. Many of these are naturally occurring substances too, like **spices**, **herbs**, **oils** and **vanilla**. Others are artificial — for example, many **esters** give artificial fruit flavourings.

Flavour enhancers don't have much of a taste themselves, but they enhance the taste of other substances. The main ones are:

1) **Sodium chloride** ('normal' salt)
2) **Monosodium glutamate**

Think about it — you've been feeding yourself all sorts of chemicals for years and years. Great, isn't it?

This table gives some examples of where different flavourings and flavour enhancers are used:

Additive	Type	Uses
herbs / spices	flavouring	wide range of savoury and sweet foods
esters	flavouring	artificial fruit flavours in sweet foods and drinks
sodium chloride	flavour enhancer	wide range of uses: soy sauce, crisps, packet noodles, etc
monosodium glutamate	flavour enhancer	

Other Additives

Preservatives *extend the* ***Shelf-life*** *of food*

Preservatives are added to food to reduce the growth of microorganisms, such as bacteria and fungi (which make food go off). Vinegar, salt and alcohol are all used for this:

1) **Vinegar** has a low pH, which **reduces the enzyme activity** in microorganisms. This means they can't metabolise properly, which stops them breaking down and spoiling the food. Vinegar is used in **pickling**.

The nicest pickled onions you have ever seen

2) **Salt** is used because the high solute concentration it creates makes the **bacterial cells lose water** by **osmosis**, so they can't grow and multiply effectively. Salt is usually used in solution as brine to preserve loads of things like tuna, olives and anchovies.

Some poisonous looking alcohol

3) **Alcohol** is **poisonous** to many microorganisms, because it is a waste product of anaerobic respiration (it's like lactate for us). Alcohol is often used to preserve fruit, e.g. pears in brandy.

Sulphites also act as preservatives. These are used to stop foods going brown (e.g. apples), but are also used in wines, beers and ciders. They act as **antioxidants** — they **reduce** the concentration of available **oxygen** in the food. But they're not all great... they can taste pretty nasty and have also been shown to reduce the concentration of the vitamin **thiamine** in foods.

Practice Questions

Q1 Name one example each of a flavour enhancer, a preservative and a colouring agent that might be used as additives in processed food.

Q2 Explain how antioxidants can stop food from going rancid.

Q3 What are the possible hazards associated with the food additive tartrazine?

Exam Question

Q1 a) Distinguish between a flavouring and a flavour enhancer, giving an example of each. [4 marks]

b) Explain how preserving works using (i) salt; (ii) vinegar; (iii) sulphites. [6 marks]

Préservatif — French for condom.......

Yes, it doesn't mean jam. Now there's a fact that could save you some embarrassment at the breakfast table on your French exchange. It doesn't have much to do with biology though. Actually, I suppose it does. But it doesn't have much to do with flavourings. Actually... I think I'll shut up now. (By the way, it's only 4 pages till the end of the section now.)

Food Storage

This page is all about how fruit ripens and how to store your food — oh the excitement, it's almost too much for me.

Metabolic Processes continue in plants even After Harvesting

Fruits and vegetables contain living cells and the metabolism (chemical reactions) in these cells continues even after harvesting. This includes **respiration** (see section 1) — stored fruits take up oxygen and respire, releasing carbon dioxide. Also, fruits and vegetables lose water by **transpiration**. These processes affect how long the fruits and vegetables stay 'fresh' — tastes, texture, colour and mass are all affected.

Fruit softens and sweetens as it ripens

Ripening is a complex process but the basic effect is the fruit getting sweeter and softer. The whole point of a fruit is to attract animals to disperse the seeds within, so they've got to taste good.

1) During ripening, **ethene** is produced — this promotes the conversion of starch to sugar.
2) Genes are switched on to make enzymes that can **hydrolyse pectins** in the plant cells walls.
3) Pectins are sticky substances that make up the **middle lamellae** between cell walls.
4) As the pectins are hydrolysed cells become leaky and the whole fruit becomes **juicy**.
5) **Acids** in the fruit are also converted into **sugars**. (Sometimes **aromatic** compounds may be produced).

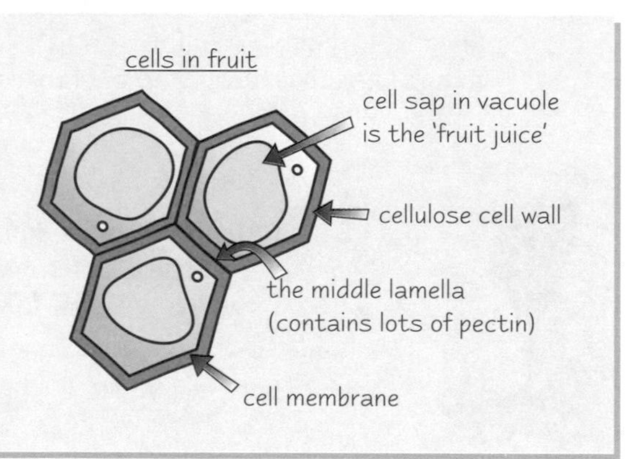

When apples ripen they become sweeter and softer. The sweetness is caused by malic acid changing into sugar. Red apples become redder - the redness of the skin is because chlorophyll breaks down and the orange-red pigments (such as carotenoids) show through. So **colour** can be a **visible indicator of ripeness**. Yum yum.

Many Processes Reduce Food Quality

The quality of food varies, depending on the levels of:

1) **Respiration** — this changes the taste of food because it **uses up sugar**, changing the balance between sugars (sweet) and acids (sour). Other substances, like vitamins, can also be affected. Respiration can make the food **smell different** too (by formation of aromatic substances).
2) **Evaporation** — water is lost and the **food becomes dry**.
3) **Softening** — this mainly affects fruits (see ripening). Softer fruits will bruise more easily.
4) **Senescence** — the **deterioration of cells**. Senescence happens when catabolic (breakdown) processes are faster than anabolic (build up) ones.

The **growth of microorganisms** in food is often the reason for reduced quality. They will rapidly multiply in the food if the conditions are right. They need:

1) **A food source** — this could be any of the nutrients (carbohydrates, proteins and so on) in the food.
2) **Suitable temperatures** — to grow and multiply.
3) **Oxygen** — (if they are aerobic).

Ripening and senescence both encourage the growth of microorganisms, speeding up decay. Moulds, for example, grow really well on the surface of ripening fruits because of the sugars released by the fruits.

Food Storage

Temperature affects Food Spoilage...

A common method to stop things going off is to **refrigerate** the food at low temperatures. This slows down the multiplication of microorganisms. Alternatively, when the food is processed it can be **heated** to kill the microorganisms present. This is normally done in one of two ways:

1) **Pasteurisation** of milk involves heating it to around 65°C for 30 minutes. This kills harmful (pathogenic) bacteria, but the milk still needs to be refrigerated to slow down the multiplication of others.
2) **Sterilisation** of milk is heating it to 100°C. This kills off more bacteria than pasteurisation (but not their spores), so it doesn't need refrigerating. This means it can be **stored for longer** than pasteurised milk before it goes off. Sterilised milk tastes funny though, because the higher temperature affects the flavour.

...as does the Storage Atmosphere

Fruits and vegetables need fresh air to respire (aerobic respiration needs oxygen, remember). So if you put an apple in an **airtight box**, the oxygen will be used up and carbon dioxide produced. This means that respiration will slow down and your apple will last for longer. You can achieve this with food packaging...

The packaging material is a **barrier** between the food and its surroundings. The best material is **partly permeable** — this minimises water loss, but stops the humidity rising to such an extent that it would encourage the growth of moulds and other microorganisms. This is called **modified atmosphere packing (MA)**. There are 2 types:

1) **Passive MA** packing lets respiratory gases through as respiration inside the container continues to take place.

2) With **active MA**, however, the air inside is **removed** and **replaced** with gas of a different composition (e.g. nitrogen) that reduces the deterioration of the food (due to respiration) as much as possible.
Sometimes **sachets of chemicals** are added that absorb the carbon dioxide or ethene.

Vacuum packaging has low permeability and keeps the oxygen concentration in the container very low.
Shrink packs are similar, except they are warmed to shrink the packaging around the food.
This reduces the growth of aerobic microorganisms, but anaerobic ones might still survive.

The **plastic films** themselves are chosen depending upon their permeability, strength, stability (under storage conditions) and transparency.

Practice Questions

Q1 Outline the changes that occur in a fruit as it ripens.

Q2 Why is there an increased risk of mould growing on ripening fruits?

Q3 What is a modified atmosphere storage system?

Exam Questions

Q1 a) Pectinase is an enzyme that hydrolyses pectins in fruits.
Explain why its use increases the yield of juice that can be extracted to make drinks. [3 marks]

b) The taste of an apple depends upon the balance between sugar (sweet) and malic acid (sour).
Explain how the processes of respiration and ripening will affect the taste of the apple. [4 marks]

Q2 Give an account of the processes of pasteurisation and sterilisation and describe how the different techniques have implications for storage of food. [8 marks]

Lunchboxes — what can I say... Linford Christie...

OK, that was a strange diversion my brain just had. Erm, yeah, lunchboxes... they're a really clever invention. They stop the bread in your sandwiches from going stale, and your apple from going brown and mushy. Just don't pick up small animals and try to take them home in your lunchbox, cos after sitting in that airtight box all afternoon, they won't be so pretty...

Microorganisms in Food

Some microorganisms are like magicians and can turn one food into another more tasty one — amazing.

Microorganisms are used to Modify foods

There are many types of foods that we produce by actually making use of the metabolic processes of microorganisms (e.g. beer, wine, bread, cheese). Some bacteria carry out **fermentation**:

In the study of **metabolism**, fermentation means the same as anaerobic respiration. However, in **biotechnology** it is used for any **metabolic process** (aerobic or anaerobic) carried out by the microorganisms. Remember that, or you'll get confused.

Fermentation often **produces acids** that **flavour** the food and **stop it going off** as quickly — by inhibiting the growth of other microorganisms.

Bacteria are used to produce Sauerkraut and Yoghurt

Making Sauerkraut

Sauerkraut is a sour-tasting food made from chopped raw **cabbage** which is allowed to ferment under anaerobic conditions in a salty environment. The cabbage loses water by osmosis and specific bacteria grow:

1) Firstly the bacteria ***Leuconostoc*** converts sugar to lactic acid until the right pH is reached.
2) Then ***Lactobacillus*** uses some of the by-products from *Leuconostoc* to produce substances that give sauerkraut its distinctive flavour.

Making Yoghurt

The conversion of milk to yoghurt also depends upon bacteria:

Lactic acid bacteria are added to either skimmed or pasteurised milk. This is heated and the bacteria clot the milk into yoghurt. This is then cooled to 5°C.

1) Some of the bacteria convert **milk protein** into **amino acids**.
2) The amino acids are then used by other bacteria to convert **lactose** in the milk into **lactic acid**.

Soy Sauce is made using Bacteria, Yeasts and Moulds

Soya beans are used to produce **tofu** (bean curd) and **soy sauce**. Tofu is made by grinding soya beans in water, filtering, and coagulating (clotting) the protein by adding vinegar.

The distinctive flavour of soy sauce is made by a combination of fermentation reactions:

1) **Starch** (as flour) is added to ground boiled soya beans and the mixture **fermented** by microorganisms at about 30°C for up to a year.
2) The **mould** *Aspergillus oryzae* in the mixture produces amylase and protease enzymes, which produce sugars and amino acids.
3) The **bacteria** *Bacillus* and *Lactobacillus* make lactic acid.
4) The **yeast** *Saccharomyces rouxii* makes alcohol.
5) The final sauce is filtered and pasteurised.

Harriet wished she had some soy sauce to make her plate more tasty.

I know it seems like I've gone off on one a bit, but you do need to know these details about making sauerkraut, yoghurt and soy sauce.

Microorganisms in Food

Yeast is important in **Breadmaking**...

Yeasts ferment glucose into carbon dioxide and ethanol. In **breadmaking**, starch is hydrolysed into glucose by the enzyme amylase (found in wheat). The glucose is then fermented by the yeast, releasing **carbon dioxide**, which provides the pockets of gas to make the dough rise.

Ascorbic acid is added to speed up the processing time:

1) The ascorbic acid oxidises the S-H (sulphur and hydrogen) groups in a wheat protein called gluten. (Oxidising removes the H atoms).
2) This makes **disulphide bridges** form (bonds between sulphur atoms).
3) This **strengthens** the dough structure so that it **traps the gas** more easily.

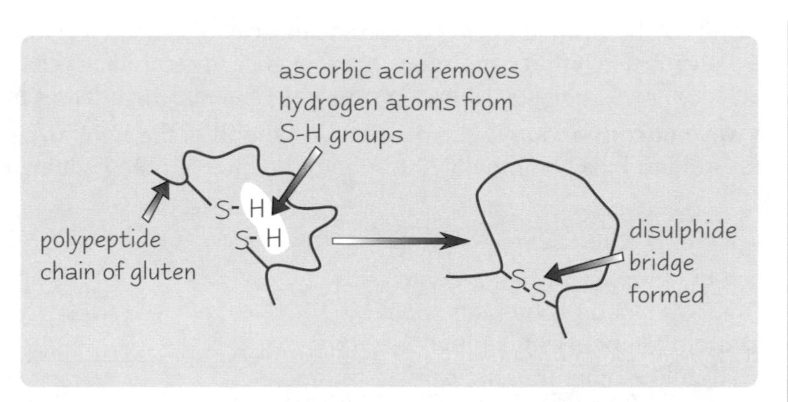

... as well as in **Winemaking**

Yeast ferments glucose in winemaking, producing **ethanol** (that's alcohol to you and me). The anaerobic respiration happens after a period of aerobic respiration. The taste of the wine depends upon the balance of **sugars** and **acids**. Black grape skins are added to make red wines and sulphites are used to prevent the growth of wild yeasts and bacteria. Different flavours develop as the wine matures. It's a pretty complicated thing and all the subtle differences mean people can prattle on about wine for ages.

Gratuitous picture of a bottle of 1968 Claret

Practice Questions

Q1 Explain how the biotechnological definition of fermentation differs from the usual metabolic version.

Q2 What is sauerkraut?

Q3 What kind of bacteria are used in yoghurt production? What do they do?

Q4 Why is ascorbic acid added to bread dough?

Exam Questions

Q1 a) Sauerkraut production requires the addition of salt and the creation of an anaerobic environment. Explain the purpose of these two factors. [2 marks]

b) Explain how the anaerobic process differs between sauerkraut production and winemaking. [2 marks]

Q2 Give an account of the role of microorganisms used in the production of soy sauce. [6 marks]

Mmmm... sauerkraut — sour, salty, fermented cabbage... sounds tasty

Microorganisms are used to make loads of things but the thought still grosses me out a little... eating yeast and bacteria by-products somehow doesn't seem natural. Well, thats the end of this section... all that talk of food has made me hungry.

The Cardiovascular System

This section covers Option C of Unit 4 of the syllabus.
The heart is a really important organ in exercise because it pumps blood to the respiring muscles.
Heart rate is controlled by a fairly complex mechanism — and I'm afraid it's part of your course...

The **Heart** is a **Pump** made of **Cardiac Muscle**

The walls of the heart are made of **cardiac muscle**. The heart has four chambers — two **atria** and two **ventricles**. These chambers **contract** and **relax** in a sequence (the **cardiac cycle**) to send blood around the body. When cardiac muscle contracts, a region of **high pressure** is created, which **forces** the blood out into a region of **lower pressure**.

The **wave of contraction** starts in a part of the **wall** of the **right atrium** called the **sinoatrial node** (SAN). Here, **sodium ions** continually diffuse into the muscle fibres, setting up **action potentials** (see p 20-21) in the fibres.

Like all muscle, cardiac muscle uses energy from ATP when it contracts. The contraction is brought about by **protein filaments** sliding over each other. But, unlike other muscle, cardiac muscle is **myogenic**, which means that it can contract **without stimulation** from the **nervous system**.

The structure of connections between the fibres in cardiac muscle means that it always contracts **smoothly** and **effectively**.

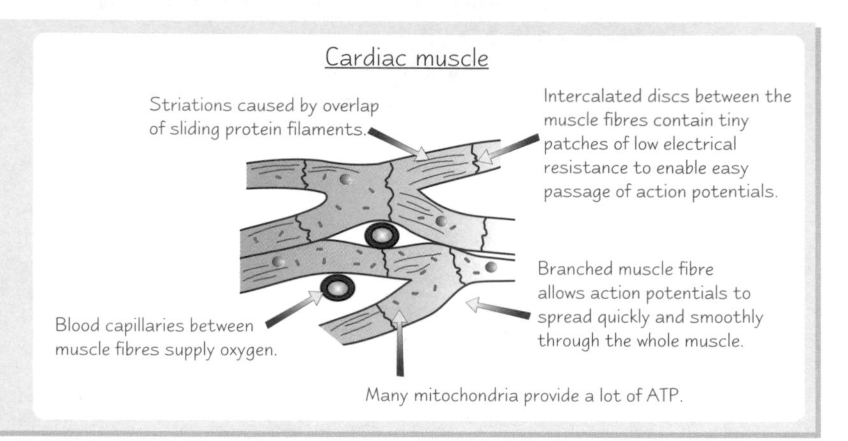

The **SAN** is Connected to the **Nervous System**

Although in theory cardiac muscle can contract on its own, its beat would be **irregular**. So in reality, it is **controlled** via **nerves** that connect to the SAN. The nerves carry **action potentials** from the **brain** which stimulate the SAN and cause the heart beat to **speed up** or **slow down** according to the body's needs.

The nerves that connect to the SAN are part of the **autonomic nervous system**. This system controls the **unconscious** activities of the body like heart rate, ventilation and digestion. The nerves are **motor nerves** carrying action potentials from the brain to the cardiac muscle. The part of the brain controlling autonomic activities is called the **medulla oblongata** (often shortened to medulla).

The activity of the **medulla** depends on action potentials being sent to it along **sensory nerves** from **receptors**. These are **'baroreceptors'** in the aorta which are stimulated when the aorta wall stretches. (See the next page for more detail.)

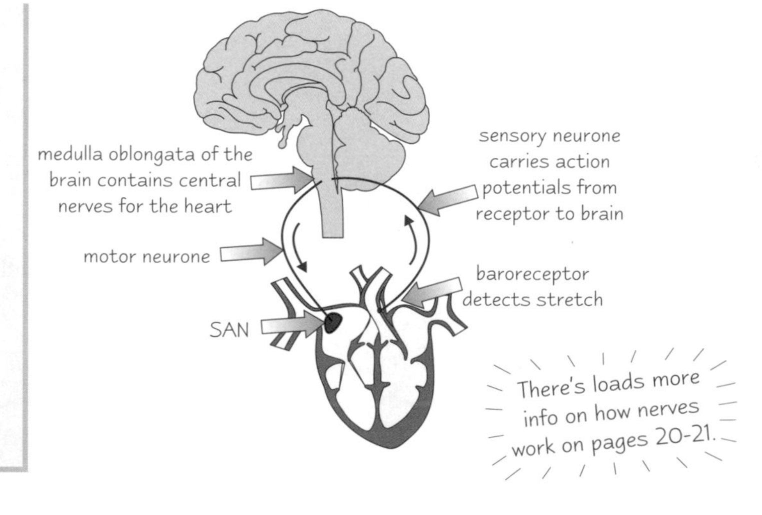

Both **Sympathetic** and **Parasympathetic** Nerves Affect **Heart Rate**

There are two kinds of **autonomic** (unconsciously controlled) nerves — **sympathetic** and **parasympathetic**. They have different effects on heart rate —

1) The brain's **medulla** contains control centres that **increase** activity in the body by sending action potentials along **sympathetic cranial nerves**. They also **decrease** it by sending them along **parasympathetic cranial nerves**.
2) The **sympathetic nerves** secrete a neurotransmitter that **increases** the rate of cardiac muscle **contraction**.
3) The **parasympathetic nerves** secrete a neurotransmitter that **relaxes** muscles and **slows down** the heart rate.

The Cardiovascular System

Autonomic Control of the Cardiac Cycle involves Reflex Responses

A **self-regulating mechanism** controls the cardiac cycle. **Baroreceptors** (stretch receptors) in the walls of certain blood vessels play an important role the cardiac cycle:

1) Blood entering the **aorta** stretches the **baroreceptors** in the aorta wall.
2) This causes an action potential to be sent through a **sensory nerve** to the **cardioinhibitory centre** in the **medulla**.
3) The medulla then sends an **impulse** along the motor **vagus** nerve (which is **parasympathetic**) to the **SAN**.
4) The neurotransmitter **acetylcholine** causes the SAN to slow the heart rate.
5) If the heart beats too **slowly**, blood **accumulates** in the **vena cava** putting pressure on the baroreceptors in the vena cava.
6) An **action potential** goes to the **cardioaccelerator centre** of the **medulla**, and then along the motor **accelerator nerve** (which is sympathetic) to the SAN. The neurotransmitter **noradrenaline** is released from the **nerve** and the heart rate increases. This increase in heart rate is called the **Bainbridge Reflex**.
7) Swellings in the carotid arteries of the neck, the **carotid sinuses**, also have **baroreceptors**. They control the **pressure** of blood flowing to the head.

The heart, carotid

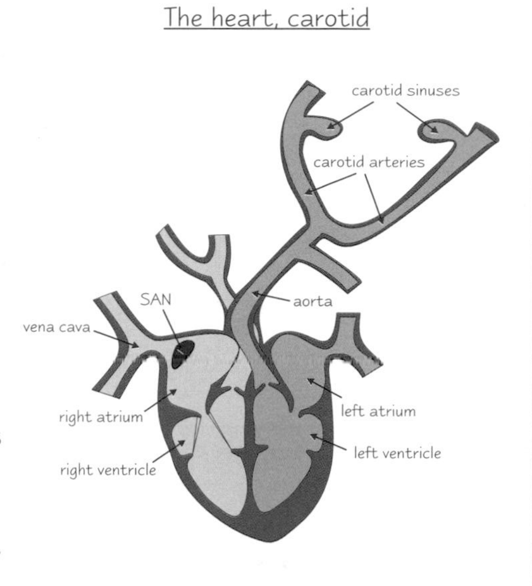

The **cardiovascular centres** of the medulla are also directly affected by substances in the blood that flows through them. During aerobic exercise the **oxygen concentration drops** and the carbon dioxide and lactate levels increase. These substances stimulate the **cardioaccelerator centre**, which brings about an increase in heart rate. Some **hormones**, such as **adrenaline** also act on the cardioaccelerator centre.

Artificial Pacemakers correct Slow Heart Beats

Sometimes **disease** of the heart can mean that the mechanisms that control heart beat **fail** to function properly. One **treatment** for this kind of disease is the use of an **artificial pacemaker**. Artificial pacemakers are **implants** near the collar bone which are connected to the vena cava. They use **electrical impulses** to stimulate the heart beat. Some pacemakers operate at a **fixed** rate whilst others detect when the heart **skips** a beat and then send a **correcting impulse**.

Practice Questions

Q1 What does the term myogenic mean?

Q2 What is a baroreceptor?

Q3 What is the difference between the sympathetic and parasympathetic nervous systems?

Q4 Explain the role of the sino-atrial node in controlling heartbeat.

Exam Questions

Q1 Describe the structure of cardiac muscle. [6 marks]

Q2 a) Dilation of the vena cava causes the heart rate to increase. Explain how this process happens, paying particular attention to the role of the autonomic nervous system. [6 marks]

b) What would be the effect of severing the nerves from the medulla to the SAN? Explain your answer. [3 marks]

Once upon a time I was falling in love, now I'm only falling apart...

...nothing I can say, just learn a couple pages on the heart. Heartbeat... why do you revise when my baby misses me... Your exam is only a heartbeat away... baby... la la la la... when it comes your way... I hope this is cheering you up after another dull page of A2 Biology facts. I can do requests. Actually, it's probably just annoying you. I'd better stop.

The Pulmonary System

Here's two lovely pages on the lungs... they're pretty important cos we'd be dead without them, so pay attention...

Lungs are specialised organs for Gaseous Exchange

Mammals exchange oxygen and carbon dioxide through their **lungs**.
You should know the structure of the lungs from GCSE and AS, but just in case your memory is a bit fuzzy, here it is again:

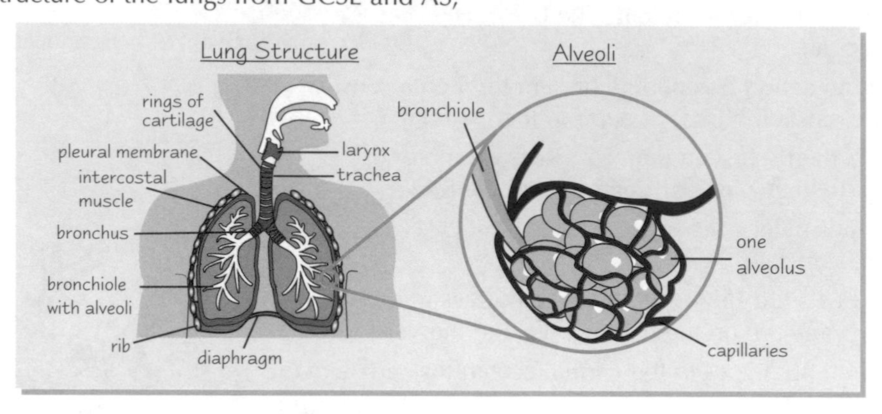

Gaseous Exchange takes place in Alveoli

Lungs are composed of millions of microscopic air-filled sacs called **alveoli**. The structure of these is specialised for the function of gas exchange. The alveoli walls are made of very thin **squamous epithelium** (that means it's made of a single layer of thin, flat cells).

There are blood-filled **capillaries** (which are also lined with squamous cells) between the alveoli. The **large surface area** of the alveoli and **short distance** between the air and the blood help maximise the rate of gaseous exchange via **diffusion**. There is also a **steep concentration gradient** of O_2 and CO_2 between the alveoli and the capillaries surrounding them, which increases gaseous exchange.

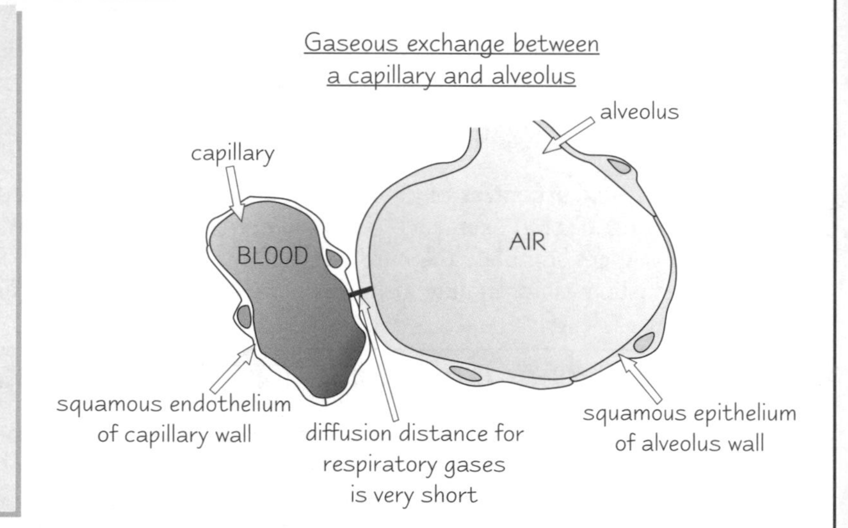

Muscles Contract and Relax to allow Ventilation (Breathing)

1) **Inspiration** (breathing in) is due to two things — **contraction** of the **diaphragm** (which flattens it down, **increasing** the **volume** of the thoracic cavity), and **contraction** of the **external intercostal muscles** (which raises the rib cage and also **increases** the **volume** of the thoracic cavity). Increasing the volume of the thoracic cavity results in a **lower air pressure** in the lungs than in the surrounding atmosphere, so **air enters** the lungs.
2) **Expiration** (breathing out) involves **relaxation** of the **diaphragm** and **external** intercostal muscles.
3) Remember that air always flows from regions of **high pressure** to regions of **lower pressure** — just like the flow of blood in the heart.

Chemoreceptors Detect Chemical Changes in the Blood

1) During exercise, CO_2 levels rise and this decreases the **pH** of the blood.
2) **Chemoreceptors** are sensitive to these chemical changes in the blood. They're found in the **medulla oblongata**, in **aortic bodies** (in the aorta), and in **carotid bodies** (in the carotid arteries carrying blood to the head).
3) If the chemoreceptors **detect** a **decrease** in the **pH** of the blood, they send a **signal** to the **medulla** to send more frequent nerve impulses to the intercostal muscles and diaphragm. This **increases** the **rate** of **breathing** and the **depth** of breathing.
4) This allows **gaseous exchange** to **speed up**. CO_2 levels drop and the demand for extra O_2 by the muscles is met.
5) A drop in **oxygen** concentration of the blood has **little effect** on ventilation.

The Pulmonary System

The *Medulla* contains *Control Centres* for *Ventilation*

The muscles which bring about **ventilation** cannot work without the nervous system.

1) The **diaphragm** and **intercostal muscles** only contract if they receive **action potentials** from the **medulla**.
2) There are centres in the medulla that control the basic breathing **rhythm** and connect to the muscles by motor nerves of the **autonomic nervous system.**
3) The **inspiratory centre** stimulates **contraction** of the **diaphragm** and **external intercostal muscles** and **inhibits** the **expiratory centre.**
4) When inspiration is complete the inspiratory centre **stops inhibiting** the expiratory centre.
5) When the **expiratory centre** operates it **inhibits** the **inspiratory centre**, so the diaphragm and external intercostal muscles **relax.**
6) The inhibition means that both centres can't stimulate the lungs at the same time and so a **breathing rhythm** is established.

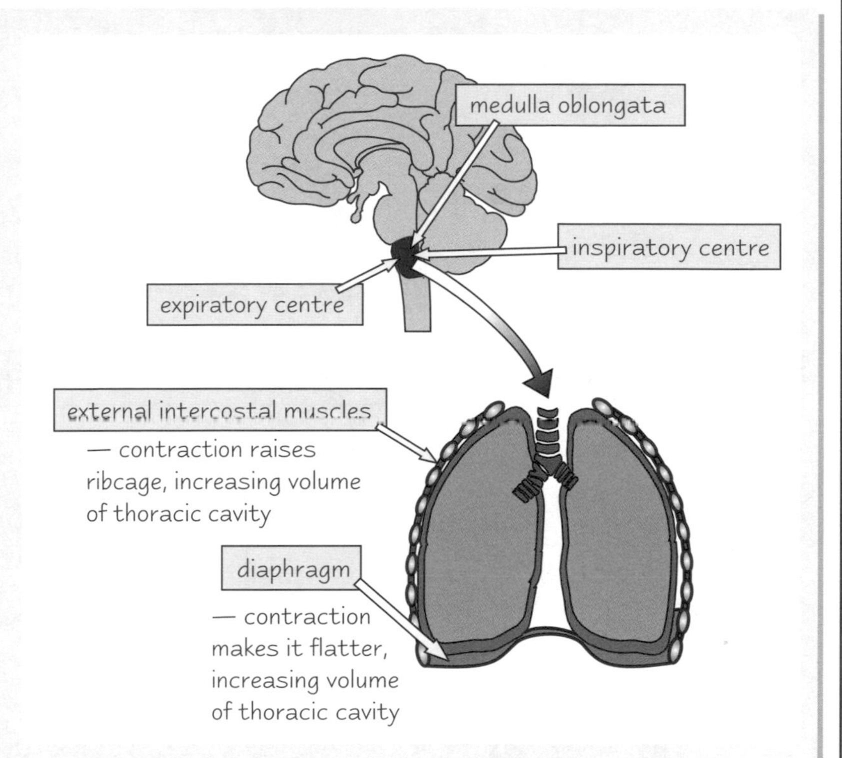

If an exam question asks you about 'cranial nerves' it just means the nerves that relay the information to and from the brain, e.g. the vagus is a cranial nerve.

Autonomic Control of Breathing involves a *Reflex Response*

There are **stretch receptors** in the walls of the **bronchi** and **bronchioles** which are stimulated when the lungs inflate during inspiration. The receptors are connected to **sensory nerves** that send action potentials along the **vagus nerve** to **inhibit** the **inspiratory centre** in the medulla. This makes inspiration stop and so expiration starts. When the lungs **deflate** the **stretch receptors** stop sending **action potentials** to the inspiratory centre so inhibition stops and inspiration commences again.

Practice Questions

Q1 Describe the structure of the lungs.

Q2 Describe the roles of the intercostal muscles and the diaphragm in inspiration.

Q3 What happens to the pH of the blood during exercise?

Q4 Explain the role of the medulla in maintaining the basic breathing rhythm.

Exam Questions

Q1 Explain how the stimulation of stretch receptors in the walls of bronchi and bronchioles causes expiration. [4 marks]

Q2 Explain how lungs are adapted to maximise gaseous exchange. [3 marks]

That medulla's at it again...

...controlling rhythms... and having centres that control things... amazing. Those two pages weren't that bad I don't think. You've probably come across a load of it before, but don't think you know it all — there's new stuff mixed in too.

The Musculo-skeletal System

Read these pages for proof that skeletons aren't that scary.

Compact Bone is made of cells called Osteocytes

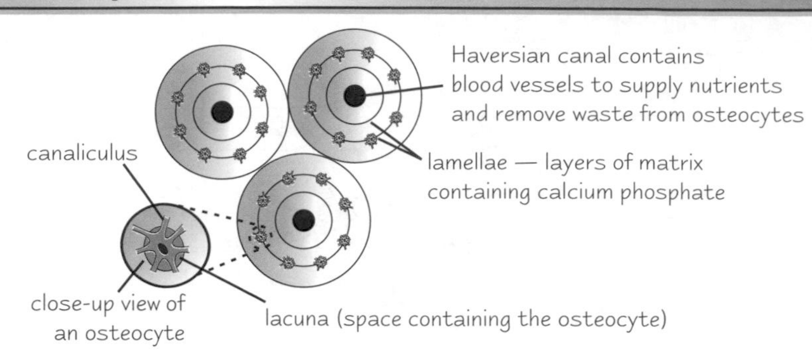

Osteocytes (bone cells) are embedded in a **matrix** of **collagen** and inorganic salts, like **calcium phosphate**. The collagen makes the bone **strong**. The osteocytes arrange themselves in **concentric rings** called **Haversian systems**, so the matrix is laid down in layers (**lamellae**) around each Haversian canal. The spaces that contain the osteocytes are called **lacunae**. The **narrow channels** crossing the lamellae are called **canaliculi**.

Synovial Joints occur at the Elbow, Knee, Shoulder and Hip

Some joints hold bones tightly together and allow **very little movement**, such as the bones of the skull, and others allow **free movement**, e.g. synovial joints.

Synovial joints act as **levers** with **antagonistic** pairs of **muscles** (e.g. the biceps and triceps in the arm). They occur where **two bones meet**.

A ball and socket joint (e.g. hip / shoulder)

synovial membrane — secretes fluid
synovial fluid — acts as a lubricant, reducing friction
cartilage — acts as 'shock absorber' and reduces friction
ligament — attaches two bones together
spongy bone — makes the skeleton lighter
rigid bone
bone marrow — where new blood cells are made

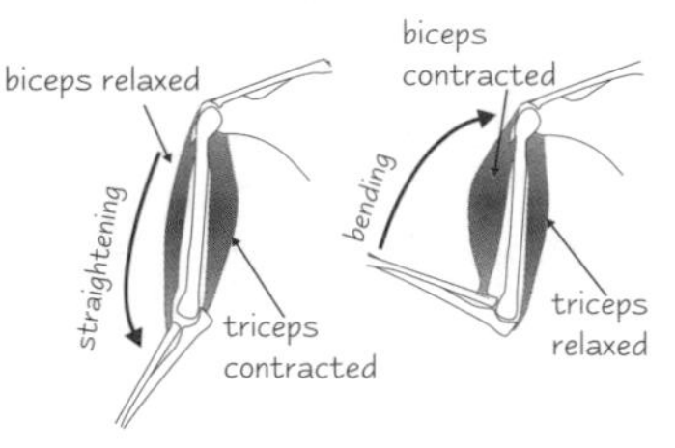

Muscles like this, that control bone movement are called **striated muscle** (also called **skeletal** or **voluntary muscle**). They are attached to bones by **inelastic tendons**. The tendons can't stretch, so when a **muscle contracts** it **shortens** and **pulls** the bone. A different muscle returns the bone to its original position.

Striated Muscle is made up of Muscle Fibres

Each **muscle fibre** contains lots of nuclei and is bound by a cell membrane called a **sarcolemma**. Contraction depends on protein **myofilaments**, which are arranged in bundles called **myofibrils**. The pattern of **thin** myofilaments (made of the protein **actin**) and **thick** myofilaments (made of the protein **myosin**) gives the **striations** (stripes) in the muscle.

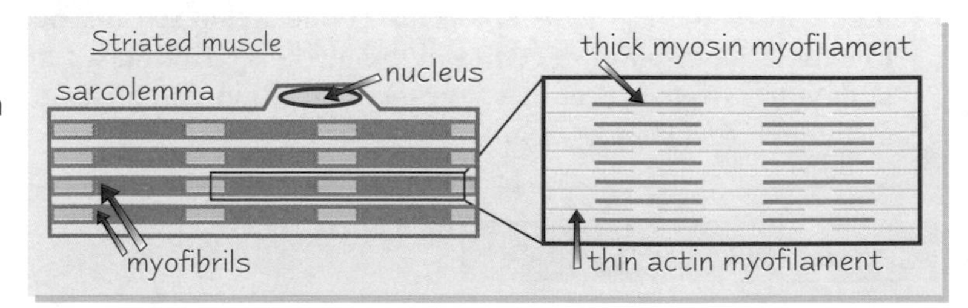

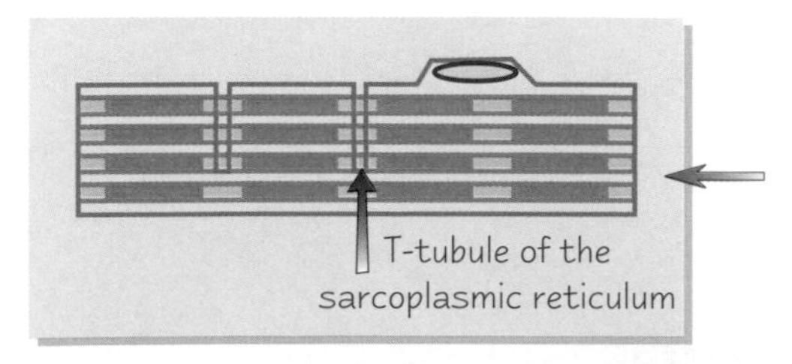

The cytoplasm of muscle fibres is called **sarcoplasm** and it's penetrated by **transverse (T) tubules** which make up a network called the **sarcoplasmic reticulum**. They let the sarcolemma transmit action potentials in towards the myofilaments. There are lots of **mitochondria** to provide ATP for contraction.

Striated Muscle is controlled by the Nervous System

Motor neurones carry action potentials to **muscle fibres**. There's a **synapse** between the presynaptic membrane of the motor axon and the postsynaptic membrane of the muscle fibre (the **sarcolemma**). This region is called the **neuromuscular junction**. The synapse functions in the same way as a synapse between two neurones and an action potential is generated on the sarcolemma in the same way too. What happens afterwards to bring about **muscle contraction** is described on the opposite page.

The Musculo-skeletal System

Muscles Contract when **Myofilaments** slide over one another

Muscle contraction is explained by **Huxley's sliding filament hypothesis**. It depends on bits of myosin called '**myosin heads**' binding to sites on actin filaments:

1) When muscle is stimulated it sets up an **action potential** in the **sarcolemma**, which spreads down the membranes of the **T-tubules**.
2) The **sarcoplasmic reticulum** membranes now become much more **permeable** and **calcium ions** diffuse out rapidly.
3) The Ca^{2+} quickly reaches the **actin filaments** and binds to a protein called **troponin**. This causes another protein called **tropomyosin** to change position and **unblock** the **binding sites** on the actin filaments.
4) ATP (from the myosin head) is hydrolysed and the energy released causes the myosin heads to alter their **angle** and attach to the binding sites forming **actomyosin cross bridges** between the two filaments.
5) The myosin head then **changes angle**, pulling the actin over the myosin towards the **centre** of the sarcomere.
6) ATP provides energy for the cross bridges to detach and then reattach, this time **further along** the actin filament. In effect the myosin heads '**walk**' along the actin filaments until they reach the end.

The process keeps repeating so that the whole muscle **contracts**.

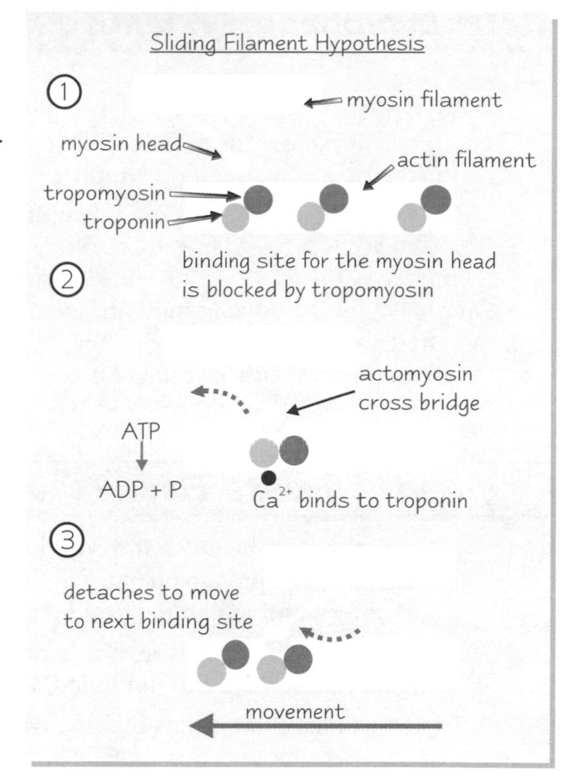

Muscles contain **Fast** and **Slow** Fibres

Fast (twitch) fibres release energy from **glycogen** and rely less on glucose delivered by the **blood**. They contain **less myoglobin** (a red pigment similar to haemoglobin) and so are white, have **fewer mitochondria** and **fatigue** more quickly. Contraction is **faster**.

Slow (tonic) fibres use **glucose** from the **blood**, have **more myoglobin** (so are red) and have **more mitochondria**. Their **slower** contractions are concerned with things like **maintaining posture**, involving muscles that are contracted for **longer periods**.

Practice Questions

Q1 Describe the structure of compact bone.

Q2 Give three examples of a synovial joint.

Q3 Explain the process of muscle contraction.

Q4 Describe the differences between fast and slow muscle fibres.

Exam Question

Q1 Describe the sequence of events that lead to the contraction of muscle following the arrival of an action potential along the T tubule. [10 marks]

Q2 Predators have many muscle fibres with particularly well developed sarcoplasmic reticulum. Suggest the advantage of this. [5 marks]

Make sure you've learnt the bare bones of this section...

This stuff isn't very difficult, but there's a lot of info crammed on to these two pages. It might be stuff that you haven't really covered in detail before, so make sure that you know it. You need to be able to recognise the structures of compact bone, synovial joints and striated muscle, so learn the main features of each. Ah go on, it'll be worth it.

The Lymphatic and Immune Systems

The lymphatic system's all about draining excess fluid from tissues and the immune system's all about protecting against disease... easy.

The **Lymphatic System Drains** *the* **Tissues**

The lymphatic system acts as a kind of drainage system that removes **excess tissue fluid**. The colourless liquid that is drained away is called **lymph**.

The system is made up of special **lymphatic capillaries** that penetrate all the tissues of the body. The lymphatic capillaries join together to form larger lymphatic vessels which empty into two large **ducts** in the **chest** cavity. From here, the lymph is returned to the **blood** via two **subclavian veins**.

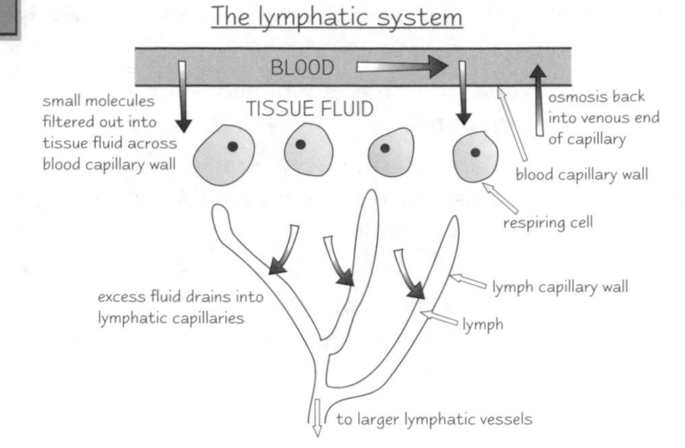

Lymph *is Excess* **Tissue Fluid** *with added* **Fatty Acids**

1) The **tissue fluid** between the cells is formed by **ultrafiltration** of the blood — cells and large molecules like proteins remain in the blood as it passes through blood capillaries, but **water** and other **small molecules** and **ions** squeeze out to form **tissue fluid**.
2) **Exchange** of substances occurs between tissue fluid and cells and lots of the tissue fluid is **reabsorbed** into the **blood**, at the **venous** end of the capillary, due to **osmosis**.
3) **Excess** tissue fluid **drains** into the **lymphatic** capillaries because of the **pressure gradient**.
4) The lymph vessels in the **villi** of the **small intestine** are called **lacteals** — the lacteals absorb **fatty acids** from digested food and these enter the lymphatic system.

The **Lymphatic System** *and the* **Immune System** *are* **Linked**

Some lymph vessels connect to clusters of tiny chambers that make up **lymph nodes**. The lymph nodes are the only part of the lymphatic system to also be included in the immune system.

The immune system is made up of loads of types of **leucocytes** (white blood cells) that defend the body against pathogens (microorganisms that cause disease). These leucocytes are all produced in the **bone marrow**.

There are five main types of leucocyte:

1) **Neutrophils** (about 70% of leucocytes) are **phagocytic** (see opposite page).
2) 24% of leucocytes are **Lymphocytes**. There are two kinds of lymphocyte:
   - **B-lymphocytes** are involved in **antibody** production.
   - **T-lymphocytes** are involved in cell-mediated immune responses.
3) **Monocytes** (4%) are **phagocytic**.
4) **Eosinophils** (1.5%) defend the body against **parasite** infection.
5) **Basophils** (0.5%) are involved in **allergy** responses.

The majority of **lymphocytes** in the body are found in the **lymph nodes**. As lymph fluid flows through the lymph nodes these white blood cells can detect the presence of **antigens** (**foreign molecules** — usually **proteins** on the surface of pathogens). This process **activates** the lymphocyte, bringing about an appropriate **immune response** (see opposite page).

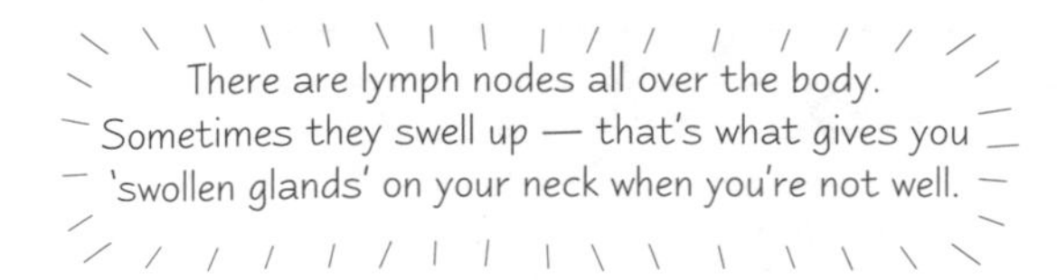

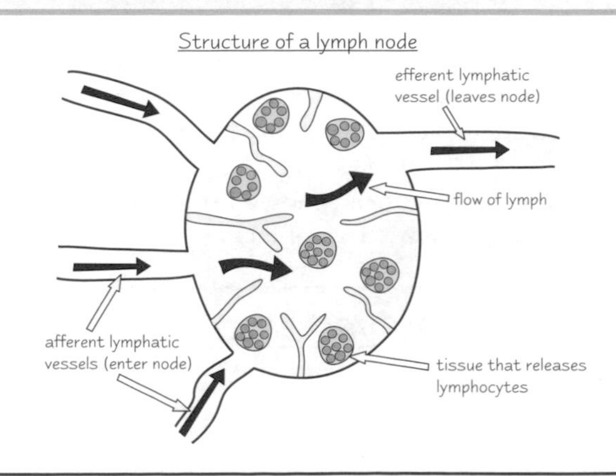

The Lymphatic and Immune Systems

Lymphocytes *have* ***Various*** *Immune Responses*

Each type of lymphocyte attacks a **specific** type of foreign particle (**antigen**). Lymphocytes belong to several classes:

1) **Killer T-lymphocytes (cytotoxic T-lymphocytes)** bind to cells (e.g. those infected by viruses) and kill them by **lysis**.
2) **Helper T-lymphocytes** stimulate **B-lymphocytes** to **divide** into antibody-producing plasma cells. B-lymphocytes can't work without helper T-lymphocytes.
3) **Plasma B-lymphocytes** secrete **antibodies** into blood plasma.
4) **Memory B and T-lymphocytes** remain in the blood for many years and produce a more effective **secondary immune response** to an infection, which is the basis for **long-term immunity**.

This is what happens when an antigen is recognised:

1) When a B-lymphocyte recognises an antigen the cell starts **dividing rapidly** by mitosis (called **clonal expansion**). This produces lots of **plasma B-cells** and **memory cells** of that type. The plasma cells then start producing lots of the **antibody** needed to fight the disease. The antibody can make the antigen harmless in the following ways:
   - **Stimulating phagocytosis** (see below).
   - **Agglutination** — antibodies make the antigens (and so the pathogens) clump together, rendering them harmless.
   - **Preventing attachment** of the pathogen to host cells.
2) When the antigen has been destroyed, the **memory cells** remain in the body. If that antigen appears again, the memory cells produce antibodies straight away, to tackle the infection. So the memory cells give you **immunity** to that disease.

This type of natural immunity, where you make your own antibodies, is called **active immunity**.

The other main type of immunity is **passive immunity**, where you get your antibodies from another organism. Babies are **passively immune** to a number of diseases when they're born, because some of their **mother's** antibodies have passed across the placenta or through her breast milk, to the baby. The passive immunity protects the baby in its first few months, and gives it a chance to build up its own **active immunity** to common pathogens.
Antibody injections are used in situations where there isn't time to build up active immunity, e.g. after a snake bite. The antibody neutralises the venom, making it harmless. This is also a passive immunity.

Phagocytosis *is where* ***Phagocytes*** *Destroy Invading* ***Pathogens***

Some white blood cells, **neutrophils** and **monocytes**, are **phagocytic**. Here's what they do:

1) Infection causes tissue inflammation. This means that blood capillaries become dilated and **leaky**, allowing **phagocytes** to leave the blood and reach the **site** of infection. (Phagocytes squeezing through capillary wall pores is known as **diapedesis**.)
2) Phagocytes **engulf** microorganisms like bacteria. The bacteria are then killed and broken down by **hydrolytic enzymes**, **free radicals** and **hydrogen peroxide**.

Practice Questions

Q1 What is the role of the lymphatic system?

Q2 Explain the difference between active and passive immunity.

Q3 Describe the process of phagocytosis.

Exam Questions

Q1 Explain the role of the lymphatic system in fighting disease. [3 marks]

Q2 Give an example of passive immunity. [2 marks]

Apparently, the 24th of January is the most depressing day of the year...

So, provided you're not revising this on the 24th of January, you've got no reason not to be joyous at the thought of learning the lymphatic system. Don't forget to learn about the immune system too.

Exercise and the Cardiovascular System

When you do exercise your heart rate increases. This page goes into a bit more detail about how the increased heart rate actually helps the body to maintain increased levels of activity. Read on my friend...

Increased Muscular Activity Leads to *Increased Oxygen Demand*

Exercise involves lots of **muscular contraction** which leads to an **increased** demand for energy. The **extra** energy needs to be produced by the **oxidation** of **glucose** in **respiration** — and the oxygen needed for this has to be supplied by the **cardiovascular** system.

Oxygen is carried in the blood bound to **haemoglobin** inside **erythrocytes** (red blood cells). Haemoglobin **loads up** with oxygen in the **lungs** (where it has a **high** affinity for oxygen) and **offloads** it in the **tissues** (where it has a **lower** affinity for oxygen).

The **oxygen dissociation curve** shows how the willingness of haemoglobin to **combine** with **oxygen** varies depending on the partial pressure (concentration) of oxygen.

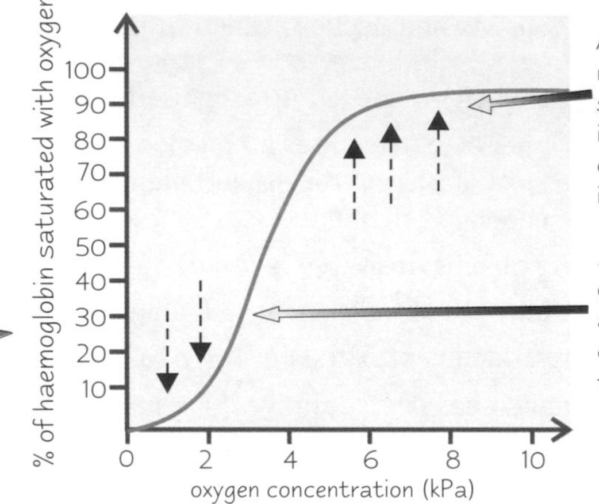

An S-shaped curve means that % saturation is improved at high oxygen concentrations in the lungs...

...and that release of oxygen is improved at low oxygen concentrations in the muscles.

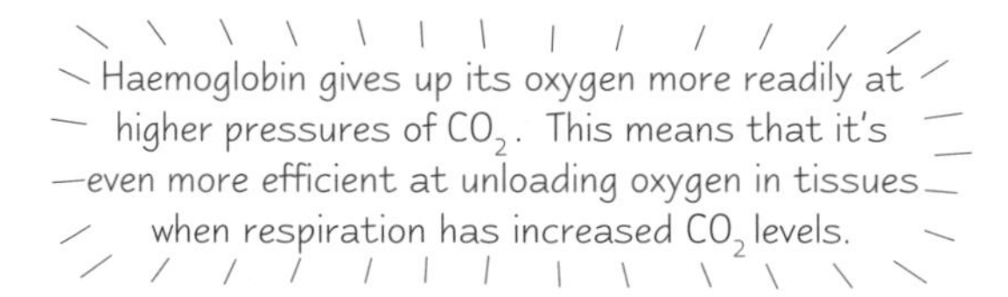

Erythrocytes (red blood cells) have several features which make them suitable for **carrying oxygen**:

1) They're packed full of **haemoglobin**.
2) They **lack** a **nucleus, mitochondria, endoplasmic reticulum** and **Golgi apparatus** — this allows more **room** to carry oxygen.
3) They are **biconcave** in shape, which maximises the **surface area: volume ratio** for absorption of oxygen. The shape also allows them to be **flexible** enough to squeeze through capillaries.

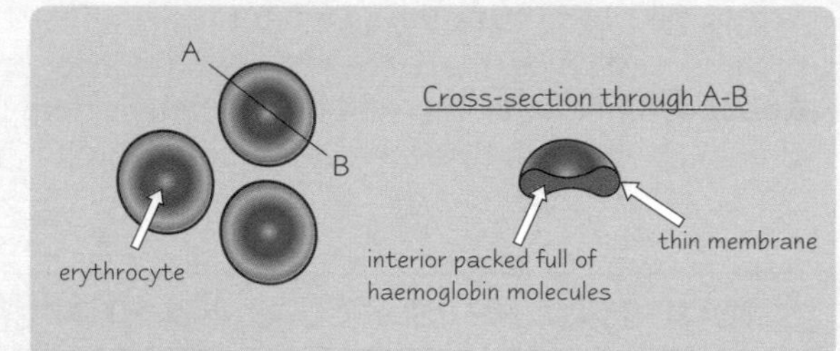

Myoglobin Stores Oxygen *in* Muscles

Myoglobin is a **blood pigment** that **stores oxygen** in **muscle tissues**. It has an **even greater affinity** for oxygen than haemoglobin. This means it acts as an oxygen **store** which only unloads when the oxygen concentration gets **really low**. When muscles are engaged in **vigorous exercise**, myoglobin releases its oxygen because the increased respiration rate quickly **uses up** the oxygen that is being provided by **haemoglobin** leaving an **oxygen deficit**.

Cardiac Output is a *Measure* of *Oxygen Delivery* to Cells

When **demand** for **oxygen increases** (e.g. during exercise) the **heart** has to deliver **more oxygenated blood** to the respiring cells. More blood (and therefore more oxygen) can be delivered if the heart beats **faster** and if each beat is **stronger**.

Stroke volume is the volume of blood pumped by each **ventricle** each time it **contracts**. The **product** of **heart beat rate** and **stroke volume** is the **cardiac output, Q**.

Q = stroke volume in cm^3 x **heart beat rate** in beats per minute

Cardiac output increases when you exercise.

So the cardiac output is the total **volume** of blood pumped by a **ventricle** every **minute** — **cm^3 per minute**. Q gives a good indication of the **effectiveness** of delivery of blood (and therefore oxygen) to respiring cells.

Exercise and the Cardiovascular System

Blood Pumped **To** Respiring Cells has to be **Returned**

In order to maintain cardiac output, blood must be **returned** to the heart by the **veins** at the same **rate** that it is pumped out. This **venous return** needs assistance because there is **low blood pressure** in the **veins**. Also, blood needs to be returned to the heart from the lower limbs against the force of **gravity**.

Venous return is helped by:

1) **Relaxation** of **heart muscle** (diastole) makes the volume of the ventricles bigger. This **reduces** the **pressure** in the ventricles so that it drops below the pressure in the veins. Blood then flows from the veins into the heart because of the **pressure gradient**.

The right side of the heart

vena cava
direction of blood flow
atrium relaxed
right ventricle relaxed

2) **Skeletal muscle** helps pump blood **back** to the **heart**. When skeletal muscles **contract** they **squeeze** the veins that run between them **increasing** the **venous pressure**. The high pressure forces the blood to flow away **towards the heart**. When the muscles **relax** the veins **fill up** and the process can begin again.

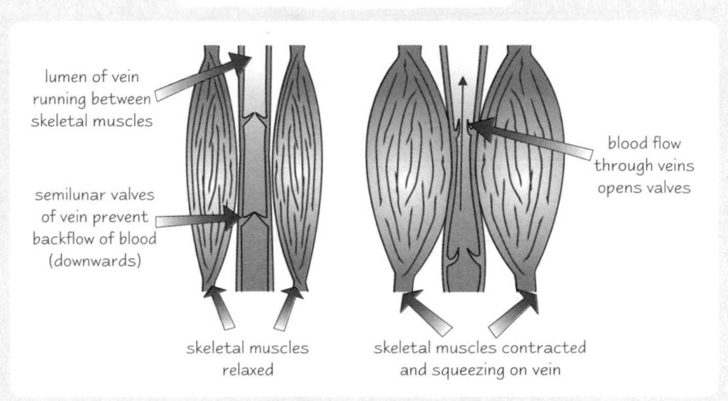

There are many **valves** in the veins. They prevent **backflow** (e.g. due to gravity). Valves themselves **cannot contract** — i.e. they can't make blood flow like the muscles can, but they can stop it flowing backwards.

Practice Questions

Q1 Explain why contraction of muscles needs oxygen.

Q2 Describe how oxygen is transported round the body.

Q3 Define cardiac output.

Q4 Explain how skeletal muscles help pump blood back to the heart.

Q5 What is the purpose of valves in veins?

Exam Questions

Q1 The stroke volume of an athlete was 100 cm^3. Her heart beat rate was 48 beats per minute. Calculate her cardiac output, showing your working. [2 marks]

Q2 What is the purpose of myoglobin in muscles? [3 marks]

Surely the best exercise and the cardiovascular system page in the world...

...ever. Anyway, that oxygen dissociation stuff is a bit tricky, so if you're stuck have a look back at your AS notes. Make sure that you can use the calculation for stroke volume — you may need it in your exam. Also, don't forget about myoglobin — there's not too much to learn but it is easy to think you know it when all you've really done is a quick skim-read.

Exercise and the Pulmonary System

Exercise doesn't just affect the cardiovascular system. It causes changes in the pulmonary system too. (If you were wondering, pulmonary means 'to do with the lungs'.)

Tidal Volume increases during Exercise

1) The volume of air **expired** during normal **ventilation** (breathing) is called the **tidal volume**. The tidal volume of an average person **at rest** is about **500 cm³**.
2) If someone breathes in as much as they can and then breathes out as much as they can, the volume **expired** is called the **vital capacity**. It can be as high as **5000 cm³**. The volume of this air which is **more** than the person **usually** breathes in and out, is called the **inspiratory** (and **expiratory**) **reserve** — it's only used when there is an **increased demand** for **oxygen**, e.g. during exercise.
3) Even with the greatest possible expiration, there is **always** some air left in the lungs (about **1200 cm³**) — this air is called the **residual volume**. This air ensures that the **alveoli** and **airways** always remain **open**.

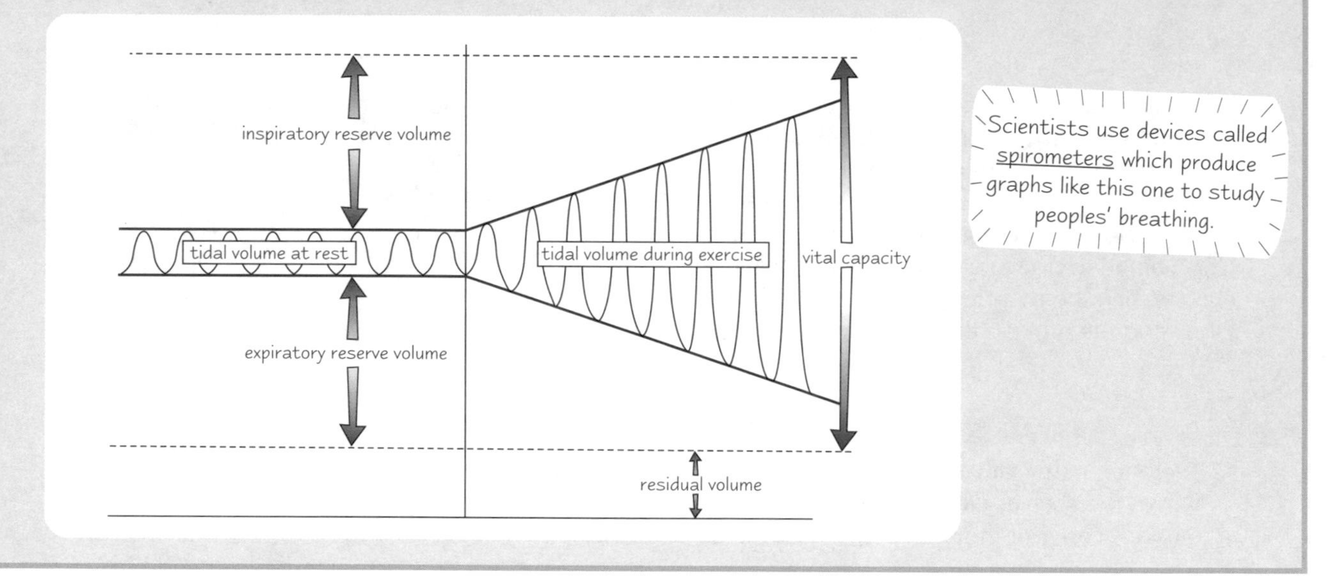

So, when you exercise you increase your tidal volume and breathing rate, but your residual volume stays the same.

Minute Volume is a Measure of Air Replacement

More **oxygen** can **diffuse** into the blood if the volume of air that is taken in during ventilation is increased. If air stays in the lungs for some time, its oxygen concentration begins to decrease because it is continually diffusing into the blood. This means that the **concentration gradient** of oxygen between the air in the **alveoli** and the **blood** becomes less steep, so oxygen diffuses more **slowly**.

So, the **more often** air is breathed in and out, the **higher** the concentration gradient, and the **more** oxygen diffuses.

During **exercise**, the **tidal volume** increases, as does the **rate of ventilation**...

The **minute volume (VE)** is the product of the tidal volume and the rate of ventilation:

VE = ventilation rate (breaths per minute) × **tidal volume** (cm³)

The **minute volume** is basically the **volume** of air breathed in and out **per minute**.
It **increases** during **exercise** — the **tidal volume** increases first and then the **ventilation rate** increases.

Exercise and the Pulmonary System

Training Increases the Effectiveness of Gaseous Exchange

Training (i.e. regular exercise) has some specific **effects** on the **pulmonary system**:

1) The **muscles** used for **ventilation** get **stronger** with **training**. This means that the **minute volume** (**tidal volume × ventilation rate**) increases. The increase in minute volume means that there is a **more effective** replacement of air in the alveoli during ventilation. This keeps the **concentration gradient** between **oxygen** in the **alveoli** and oxygen in the **blood** as **high** as possible.

2) More **capillaries** develop around the **alveoli** which increases the amount of **blood** available for **oxygen diffusion**.

3) In combination, this **improved ventilation** and **better blood flow** makes the concentration gradient of respiratory gases between the air in the alveoli and the blood in the capillaries **steeper**. This means they diffuse **faster**.

4) Remember that getting rid of **carbon dioxide** is just as important as getting in oxygen. The increases in blood flow to the alveoli and higher turn-over of air in the lungs brought about by **training** means that the **concentration** of carbon dioxide in the **blood** is much **higher** than in the alveoli so it can be **efficiently** expelled via **diffusion from** the **blood into** the **air** in the lungs.

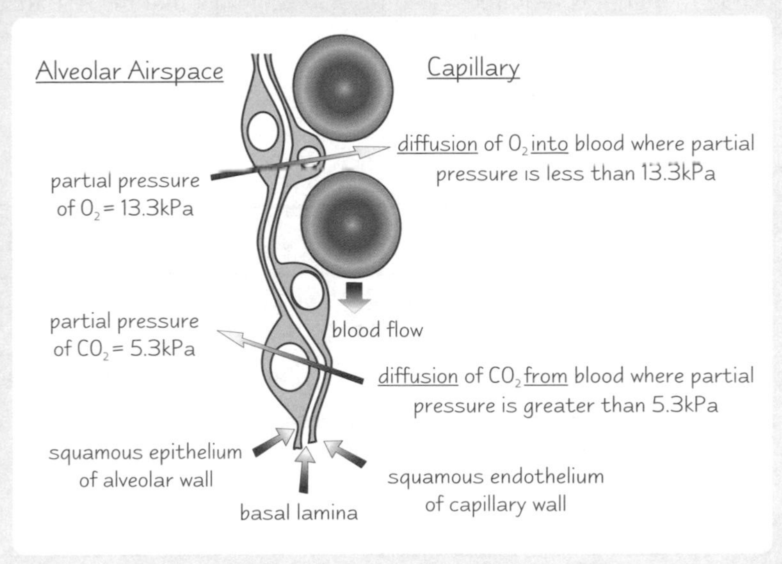

Ventilation requires Energy

Ventilation involves **contraction** of the **intercostal muscles** and the **diaphragm**, especially during inspiration. Inspiration is therefore an **energy-demanding** process — **respiration** is needed to provide energy for muscles to contract, so ventilation uses **oxygen** and **glucose**.

Practice Questions

Q1 What is the difference between tidal volume and vital capacity?

Q2 What effect does an increase in ventilation rate have on the concentration gradient between air in the alveoli and blood in capillaries of the lungs?

Q3 Give a definition of 'minute volume'.

Q4 Explain why ventilation needs energy.

Exam Questions

Q1 a) Describe the effects of exercise on: (i) ventilation rate (ii) minute volume. [2 marks]
b) What is the inspiratory reserve volume? What is the effect of exercise on this measurement? [2 marks]

Q2 Describe the changes that occur in the alveolar tissues of the lungs as a result of training and explain how these changes affect gaseous exchange. [6 marks]

Perhaps the finest exercise and the pulmonary system page in the universe...

Oh, but I think it is. Well, there's not too much to learn on these pages. Basically, it all comes down to concentration gradients — training increases the body's abilities to keep O_2 and CO_2 flowing in and out of the blood efficiently. And that means that your muscles can carry on contracting for longer.

Exercise and the Musculo-skeletal System

Things are about to get complicated. Sorry. But bear with me, it's not too bad once you've got your head around it.

ATP** must be **Regenerated** in **Contracting Muscles

The **energy** for muscle contraction comes from the **breakdown** (hydrolysis) of **ATP**:

ATP → ADP + P + energy

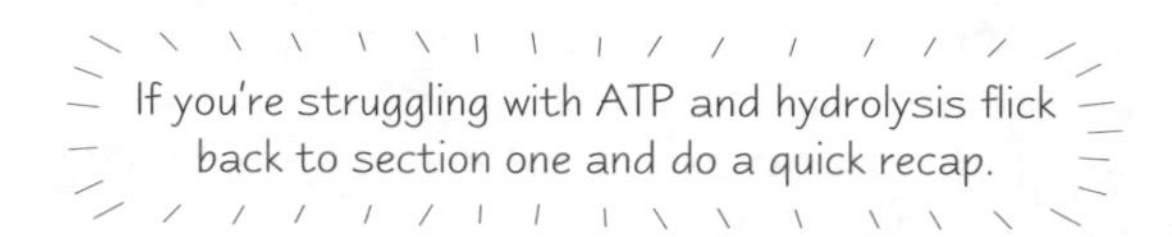

So much energy is required when muscles contract that the ATP that's available gets used up quickly. ATP has to be continually reformed so that exercise can continue — this happens in **three main ways**:

1) **ATP-PCr System** — a substance called **phosphocreatine** donates a **phosphate** group to ADP to reform ATP:

PCr (phosphocreatine) + ADP → ATP + Cr (creatine)

This is good for regenerating ATP in **short bursts** of vigorous exercise like a 100m sprint. The system is **anaerobic** (it doesn't need any oxygen) and it's **alactic** (it doesn't form any lactate).

2) **Lactic acid system** — ATP is made by glycolysis (that's conversion of glucose into **pyruvate** in the **cytoplasm**). This is a good system for longer periods of vigorous exercise e.g. a 400m sprint. The **pyruvate** is then reduced to **lactate**. The **reduction reaction** involves the addition of **hydrogen** from the **reduced NAD** formed in **glycolysis**. The lactate (**lactic acid**) may **accumulate** in the muscles and cause **muscle fatigue**. After exercise has finished, oxygen is needed to **reoxidise** the lactic acid — that's called the **oxygen debt**. Some lactate is also carried to the **liver** where it is changed to **glucose** and then stored as glycogen.

Lactic Acid Reaction

2ADP 2ATP
Reduced NAD
NAD
glucose → pyruvate → lactate (lactic acid)
Glycolysis

3) **Aerobic respiration** involves **glycolysis** as well as reactions inside the **mitochondria**. This generates a lot of ATP mostly via **oxidative phosphorylation** (see page 7). Aerobic respiration only works in the presence of **oxygen** and during prolonged periods of exercise it is usually **dominant**.

Muscles** Send **Information** to the **Brain

The **sliding filament mechanism** is how muscle contraction occurs when **muscle filaments slide** over one another (see page 59). Between these filaments are bundles of fibres of a different type called **spindles** — they're connected to **sensory receptors**. When spindles are **stretched** they trigger the firing of **action potentials** along **sensory neurones** to the **brain**.

This means that the brain constantly receives information on changes in muscle lengths all over the body and so it can respond and **coordinate** bodily **movement**.

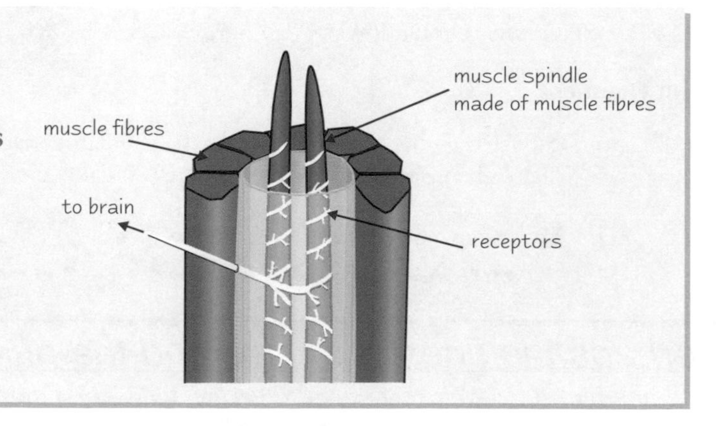

Exercise and the Musculo-skeletal System

A **Motor Neurone** with its **Connected Motor Fibres** makes up a **Motor Unit**

1) When an **action potential** arrives at the end of a motor neurone, a **neurotransmitter diffuses** across the **synapse** to create an action potential in the muscle **sarcolemma** (fibre membrane). This causes the muscle to **contract**.
2) In combination, the **motor neurone** and the **muscle fibres** are connected to make up a **motor unit**. Motor units with **few** muscle fibres per motor neurone can bring about **finely coordinated** movement, for example, movement of the **eye**.
3) When many **action potentials** arrive at muscle fibres in **quick succession** there is a **sustained contraction** called **tetanus** —it happens because the fibres do not have time to relax between contractions.

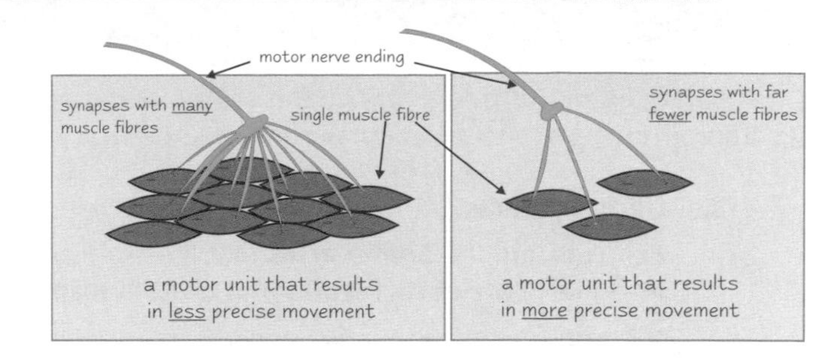

Tetanus is also the name of a disease caused by bacteria found in soil. Tetanus stops the inhibition of nerve impulses to the muscles so they all contract (i.e. they go into tetanus). Tetanus is sometimes called 'lockjaw' because the muscles that close the jaw are much stronger than those that open it so sufferers end up not being able to open their mouths. Nice.

Muscles Don't Always **Shorten** when they Contract

There are two key types of muscle contraction —

1) **Isotonic** muscle contraction happens when muscles contract and shorten whilst maintaining a constant tension.
2) **Isometric** muscle contraction is when the tension of the muscle increases but the length doesn't change.

Isometric contraction happens when your arm holds a heavy weight. It's important in maintaining **posture** too.

Muscle **Strength** Depends on its **Cross-Sectional Area**

Muscles with a large cross-sectional area generate greater **shortening forces** so they are **stronger** than narrower muscles. Wider muscles have the same number of muscle fibres, but they have more **myofibrils**, the filaments which make up a muscle fibre.

Although they generate greater forces, wider muscles fatigue quickly because of the **depletion** of **ATP** and **glycogen** and the build-up of **lactate**.

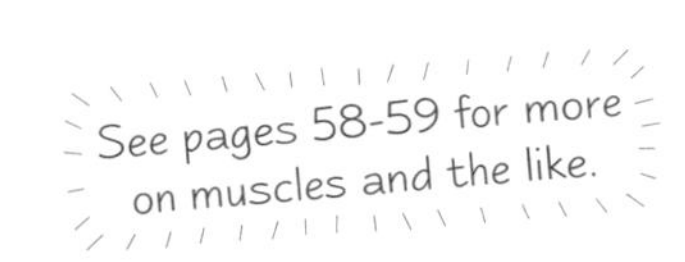

Practice Questions

Q1 Describe two anaerobic sources of ATP.

Q2 What causes muscle fatigue?

Q3 Define the following terms: (i) motor unit (ii) tetanus.

Q4 What is the difference between isometric and isotonic muscle contraction?

Exam Questions

Q1 a) Explain why lactate can build up in skeletal muscle [3 marks]
b) What is meant by the term 'oxygen debt'? [2 marks]

Q2 If creatine is taken as a dietary supplement, more short-term ATP can be made available to working muscles. Explain this effect. [3 marks]

Q3 Describe the principal sources of ATP for muscle contraction during a 10 minute period of running. [6 marks]

Just when you thought you were safe — ATP returns for seconds...

You probably thought you'd seen the last of ATP. If only. The trouble is that ATP is so central to Biology that it just keeps coming up again and again. Remember that the fundamentals of respiration are covered in section one, so you can always look at that to jog your memory. Other than that, there's only one thing that you can do — learn it.

Training

If you're into sport, or you're doing A level PE then you'll probably be familiar with some of the things on these pages. If you're not then perhaps this'll inspire you to take up marathon running.

Effective **Training** Requires a **Balanced Diet**

A **balanced diet** is one that includes all the necessary **food compounds** in the appropriate amounts to maintain **health**. The main food compounds are:

1) **Carbohydrates** and **lipids** are respired to release **energy**.
2) **Proteins** and the **amino acids** that they're made of are needed to make **enzymes**, **carrier molecules** and **muscle filaments**.
3) **Vitamins** are **compounds** needed in very small quantities for specific functions (e.g. some are coenzymes which help enzymes work).
4) **Minerals** include **ions of salts** and are needed for various functions, for example, calcium is an important constituent of bones and is needed for muscle contraction. Minerals are only needed in small quantities.
5) **Water** makes up about **80%** of cell contents. It's crucial because most **biological reactions** take place **in solution**. During exercise the body's demand for water rapidly increases.

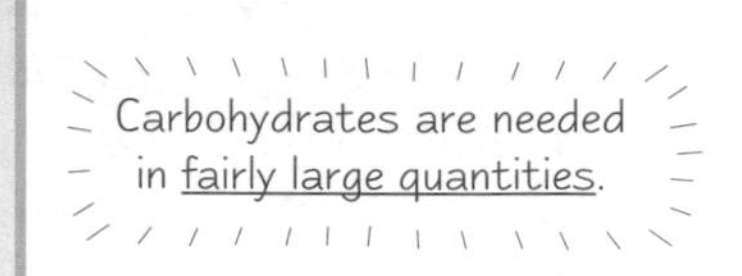

The quantity of these substances that the body needs depends on the type of activity — very strenuous exercise over a long period of time uses more fat than carbohydrate.

Anaerobic Conditioning is Important for **Short-Burst Exercise**

Training for sports that involve **brief bursts** of exercise tends to focus on **anaerobic conditioning**. Anaerobic conditioning increases the **efficiency** of ATP production via the anaerobic systems.

For example, **interval training** uses **short** (less than a minute) **bursts** of exercise separated by brief (30 second) periods of **rest**. The rest periods mean that there isn't much time for **lactate** to build up and so the production of **ATP** by the **ATP-PCr** system is **maximised**.

Aerobic Training is Important for **Prolonged Exercise**

The **aerobic system** is the main source of ATP for **long** periods of exercise.
As you know, aerobic exercise increases the demand for oxygen at the contracting muscles.
Repeated aerobic training has several **physiological consequences:**

1) **Cardiac output** increases — due to an increase in **stroke volume**.
2) **Blood flow** to the **skeletal muscle** is increased. As a result blood flow to **less active organs**, e.g. the gut, is **reduced**.
3) **Venous return** to the **heart** is **improved**.
4) The **minute volume** of the lungs increases and the number of **alveolar capillaries** increases. In combination this means that there is a greater capacity to load the blood with oxygen and offload carbon dioxide.
5) The number of **mitochondria** per **muscle fibre** increases, so there is more **aerobic production** of **ATP**.
6) Aerobic training makes the muscles get increasing amounts of their **energy** from **lipids**, rather than **carbohydrates**.
7) Muscles take longer to become **fatigued**.

Aerobic training of muscles is sometimes called endurance training.

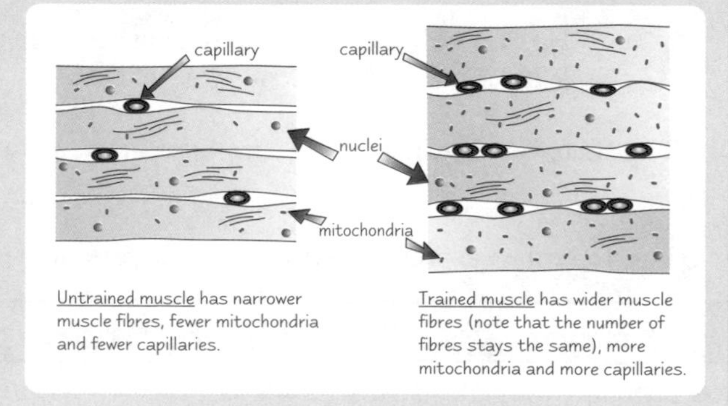

Untrained muscle has narrower muscle fibres, fewer mitochondria and fewer capillaries.

Trained muscle has wider muscle fibres (note that the number of fibres stays the same), more mitochondria and more capillaries.

Training

Training Improves Performance

Dave believes the key to achieving peak performance is wearing the right pants.

1) **Training** involves **conditioning** the body for a **particular type** of exercise.
2) Different types of exercise use different methods of **reforming ATP**.
   This is because different types of exercise have different **energy demands**.
3) Short bursts of vigorous exercise use more of the **anaerobic** systems (ATP-PCr and lactic acid systems), whereas long periods of exercise, such as marathons, are more **aerobic**.

Muscular Exhaustion is Called Fatigue

After strenuous activity, muscles can become **fatigued**.

Fatigue has several different **causes**:
1) There is a **lack** of available **ATP**.
2) **Glycogen** stores have **run out**.
3) **High** levels of **lactate** have accumulated.

During exercise, **glycogen** stores are **hydrolysed** to provide the **glucose** that is needed for **respiration**.
Athletes can increase the amount of glycogen in their muscles before they exercise by **glycogen loading**.

To do this, athletes keep their glycogen levels **low** for a few days by reducing **carbohydrate** levels in their **diet**.
When they eat carbohydrates again, the body **overcompensates** and stores large amounts of **glycogen**.
The increased glycogen store means that muscles can be worked for **longer** without fatigue being a problem.

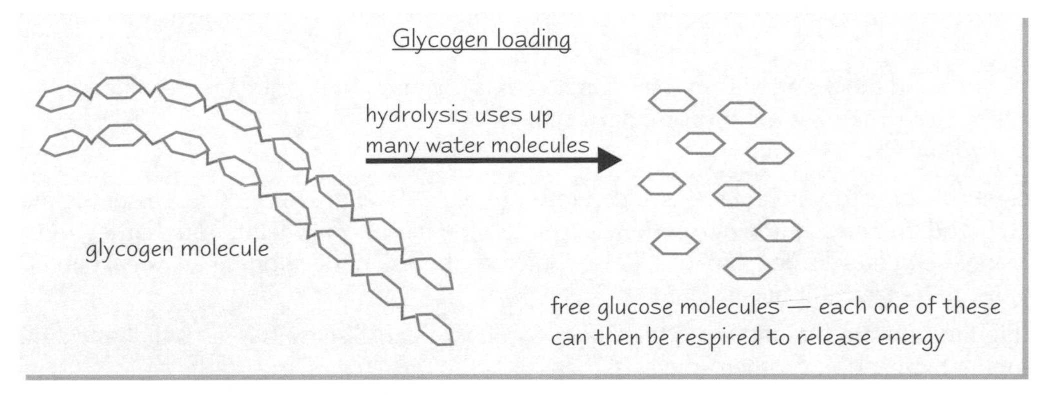

Practice Questions

Q1 Explain the importance of carbohydrates, lipids and protein in the diet of an athlete.

Q2 What role do minerals play in diet?

Q3 Explain what is meant by anaerobic conditioning.

Q4 Explain what is meant by endurance training.

Exam Questions

Q1 a) Explain the training involved for a 100m sprint. [5 marks]
b) Explain what is meant by the term 'glycogen loading'. [6 marks]

Q2 How does a trained muscle differ from an untrained muscle? [3 marks]

Ooh, I like Dave's pants...

Well, there you go, a whole page on training. I'd say that revision is more like a marathon than a sprint so you'd better do some endurance training and increase the number of capillaries in your brain. Also, before the exam you can do some glycogen loading by eating lots of chocolate and pizza — it's the best way to ensure that you do well.

Human Disorders

There's one basic rule in life — anything you enjoy is likely to be bad for your health. With the exception of exercise, but who really prefers jogging to pizza? Most things are OK in moderation, but if you overdo the fun, you pay for it. Boo.

You need to Take Care of your **Heart** and **Arteries**

Here are some of the nasty things that can happen (even the words are pretty nasty, but the effects are worse):

Atherosclerosis

This is the 'furring up' of the **coronary arteries** by fats — mostly **cholesterol**.

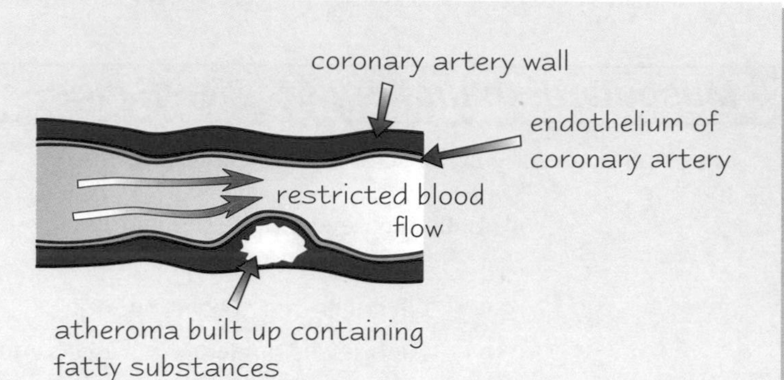

- **Fatty deposits** called **atheromas** (also called atheromatous plaques) form beneath the artery endothelium lining, **narrowing** the vessel. This reduces the blood flow as there is increased resistance.
- The fatty material mainly comes from cholesterol in the plasma, carried by **low-density lipoproteins (LDL)**.
- Atherosclerosis makes the formation of **clots** more likely. If the membrane over the plaque ruptures (breaks open) then blood cells stick to the exposed material. This forms a blood clot.

Heart Attack

If a blood clot blocks the arteries supplying part of the heart the blood supply is reduced to the tissue. This causes the heart to be damaged and results in a **myocardial infarction** (heart attack).

Stroke

If a blood clot forms and blocks arteries in the brain it causes localised brain damage, resulting in a **stroke**. The medical term for a stroke is a **cerebrovascular accident**.

Hypertension

Being overweight or eating too much fat or salt can result in **high blood pressure**. Of course, the medical term isn't 'high blood pressure' but **hypertension**. Hypertension makes your heart work harder and can increase the risk of heart attacks and strokes. It also increases the risk of developing an **aneurysm** (a bulge in an artery) which can be **fatal** if it bursts.

A diet that is high in **saturated fat** increases the risk of **coronary heart disease** (diseases that affect the arteries that supply the heart, e.g. atherosclerosis).

Osteoporosis and ***Osteoarthritis*** are diseases to do with your ***Bones***

Osteoporosis and **osteoarthritis** are two common problems with the skeletal system which often occur in **older people**:

1) **Osteoporosis** is a loss of **calcium** from **bone tissue**, which makes bones weaker, **more brittle** and **more likely to fracture**. This affects **women** more than men — calcium needed by the body comes from the bones, and it's controlled by a hormone called **parathormone**. **Oestrogen inhibits parathormone** but during the **menopause**, as oestrogen levels **decrease**, **more calcium** is **removed** from the bones. Osteoporosis can be treated with calcium supplements or, for women during the menopause, Hormone Replacement Therapy (HRT).
It can be prevented by getting lots of calcium in your diet and doing weight-bearing exercise (e.g stair climbing).

2) **Osteoarthritis** is where the **cartilage deteriorates** and wears away, and there's less **synovial fluid** in the joint. The bones can get **damaged** or deformed as they **rub** against ligaments and other bones. This can lead to painful, swollen joints. It can be treated with drugs for the pain and swelling, or in severe cases, joint replacement — that's why a lot of older people have **hip replacements**.

Human Disorders

There can be **Nasty Consequences** if you Don't Take Care of your **Lungs**

Here are some nasty diseases caused by mistreatment of your **lungs**.

1) **Lung cancer**. The main cause of lung cancer is **smoking** and you're also more likely to get it if you live or work in an atmosphere where other people smoke. **Tumours** grow in the epithelial tissue of the lungs and damage them, and tumour cells travel around the body in the blood and set up **secondary tumours** (**metastases**) elsewhere.
2) **Chronic bronchitis** is another disease that may result from **smoking**, and it can also be caused by **air pollution**. Irritation causes the bronchioles in the lungs to become **inflamed**, which can block some of them. This makes breathing more difficult.
3) **Pneumoconiosis** is fibrosis and scarring of the lungs due to long-term breathing of dusts like coal, silica or asbestos. It's seen mainly in people who work in dusty environments, like coalmines or quarries.
4) **Tuberculosis** (**TB**) is an **infectious** disease caused by a bacterium (*Mycobacterium tuberculosis*), unlike the other four. Symptoms include a bad cough, fever and weight loss. See page 27 for more on TB.

TREATMENT OF HEALTH PROBLEMS

Disease	Treatment
atherosclerosis	surgical removal of section of artery, angioplasty (unblocking of arteries using a needle)
hypertension	low sodium diet and drug treatment (e.g. diuretics, beta blockers, calcium channel blockers), low fat diet and exercise programme
coronary heart disease	surgery, low fat diet, exercise programme, drug treatment (e.g. aspirin, statins, nitrates)
osteoporosis	calcium supplements, HRT (for women during the menopause)
osteoarthritis	drug treatment (anti-inflammatory drugs, painkillers), joint replacement surgery
chronic bronchitis	drug treatment (e.g. bronchodilators, corticosteroids), oxygen therapy
tuberculosis (TB)	drug treatment (long-term course of several antibiotics)
pneumoconiosis	no treatment, except to avoid further inhalation
lung cancer	radiotherapy, chemotherapy, surgery

Practice Questions

Q1 What is an atheroma?

Q2 What is the common name for a cerebrovascular accident?

Q3 What are osteoporosis and osteoathritis?

Q4 Name two lung diseases that can be caused by smoking.

Q5 Give one type of treatment for each of the diseases you named in your answer to Q4.

Exam Questions

Q1 Explain the health problems that can be caused by a diet too high in saturated fats. [6 marks]

Q2 a) Explain why osteoporosis-related bone fractures are significantly more common in women than men by the time they reach the age of 70. [2 marks]

b) Explain how osteoarthritis causes joint pain. [2 marks]

TB or not TB? That is the question...

...hopefully not TB is the answer. These two pages aren't that bad really, but there are quite a few medical terms you have to learn, especially in that first section. Doctors like to sound clever, so they use words like 'myocardial infarction' instead of the much simpler 'heart attack', and you'll have to too.

Photosynthesis

You remember photosynthesis — carbon dioxide and water plus light energy makes tasty glucose, and the rather useful by-product of oxygen. It's very popular in plants, where it mostly tends to go on in the leaves. Sound familiar yet?

Leaves are Adapted for Photosynthesis

Leaves are **organs** — they contain groups of **tissues** that work together to carry out **photosynthesis**.

1) Leaves are usually **broad**, **thin** and **flat** — this helps them absorb as much of the available **light** as possible. It also means that **carbon dioxide** entering the leaf through the **stomata** can reach inner cells easily.
2) **Veins** branching through the leaf contain **xylem vessels** to bring **water** from the roots and **phloem vessels** to carry away the **sugars** made in photosynthesis to the rest of the plant.
3) The waxy **cuticle** covering the **upper epidermis** protects the leaf from damaging **UV rays** in the sunlight. It's also **waterproof**, which helps to prevent **dehydration**, and clear so light can penetrate it.
4) The **palisade cells** in the upper part of the leaf contain the most **chloroplasts** because they're close to the light striking the top of the leaf. Chloroplasts are where photosynthesis happens — they contain the **light-absorbing pigments** (see below).
5) **Air spaces** between the cells in the **spongy mesophyll** layer allow gases like **carbon dioxide** and **oxygen** to diffuse easily through the leaf. There are also spaces between the palisade cells for the same reason.
6) **Stomata** (pores) in the lower surface of the leaf allow exchange of carbon dioxide and oxygen between the leaf and the **atmosphere**.

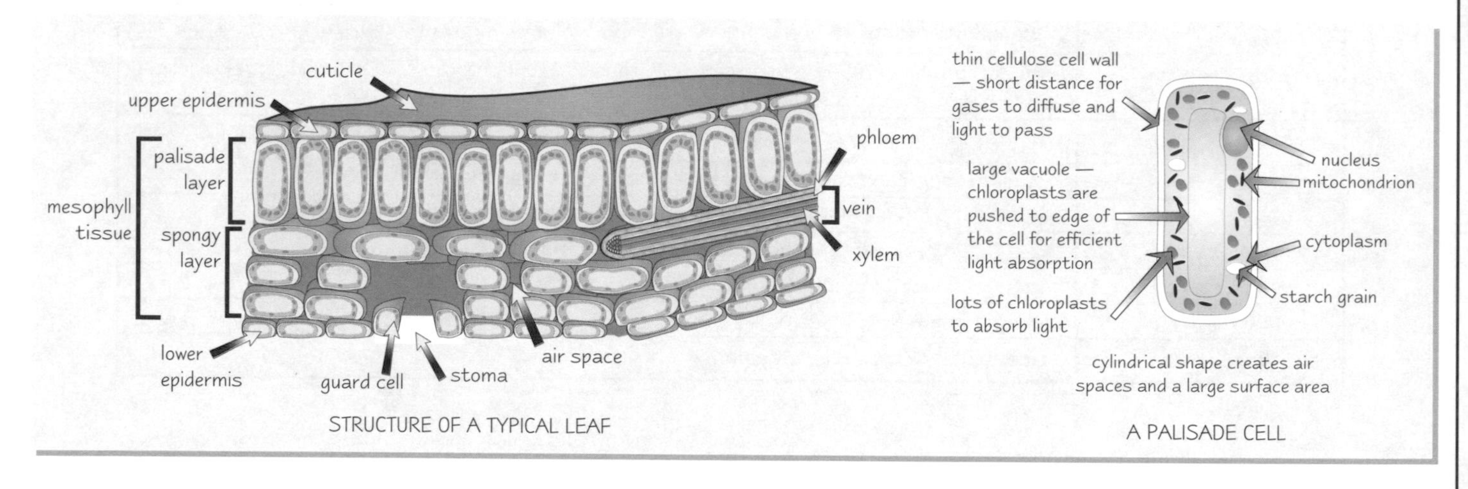

STRUCTURE OF A TYPICAL LEAF

A PALISADE CELL

Chloroplasts are the Site of Photosynthetic Reactions

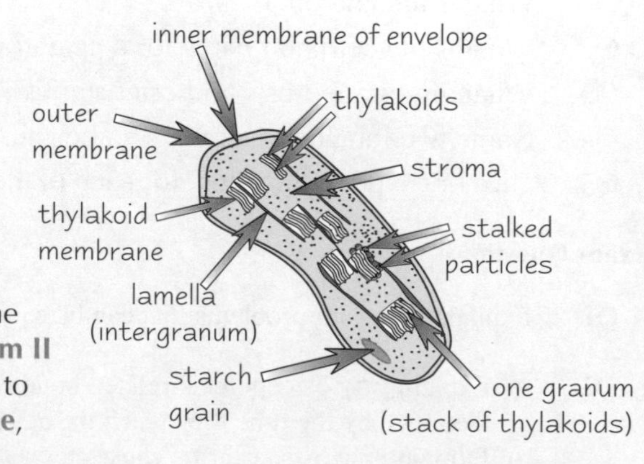

1) Chloroplasts are usually about **5 μm** in diameter.
2) They're surrounded by a double membrane called the **chloroplast envelope**.
3) **Thylakoids** (fluid-filled sacs) are stacked up inside the chloroplast into structures called **grana**. These structures have a large surface area. The thylakoids are where the **light-dependent reaction** of photosynthesis occurs.
4) **Chlorophyll** and other photosynthetic pigments are found on the **thylakoid membranes**. They form a complex called **photosystem II** (see page 76). Some thylakoids have **extensions** that join them to thylakoids in other grana. These are called **inter-granal lamellae**, and they're the sites of **photosystem I** (see page 76).
5) The thylakoids are embedded in a gel-like substance called the **stroma**. The stroma is where the **light-independent reaction** of photosynthesis (called the **Calvin cycle**) happens. It contains **enzymes** (for the Calvin cycle), **sugars** and **organic acids**.
6) Carbohydrates produced by photosynthesis and not used straight away are stored as **starch grains** in the **stroma**.

Photosynthesis

Photosynthetic Pigments absorb Visible Light

The **thylakoid membranes** inside chloroplasts contain **photosystems**, which are groups of **pigment molecules**. Three of the most important pigments found in plants are **chlorophyll a**, **chlorophyll b** and **carotene** (a carotenoid). Experiments have shown that each of these pigments absorbs a **different wavelength** (colour) of light:

1) The different pigments can be separated using **chromatography**.
2) A **spectrometer** is used to measure the absorption of light of different wavelengths by the different pigments.
3) This gives an **absorption spectrum**, showing the **absorption** of each pigment over the wavelength range of **visible light** (see diagram).

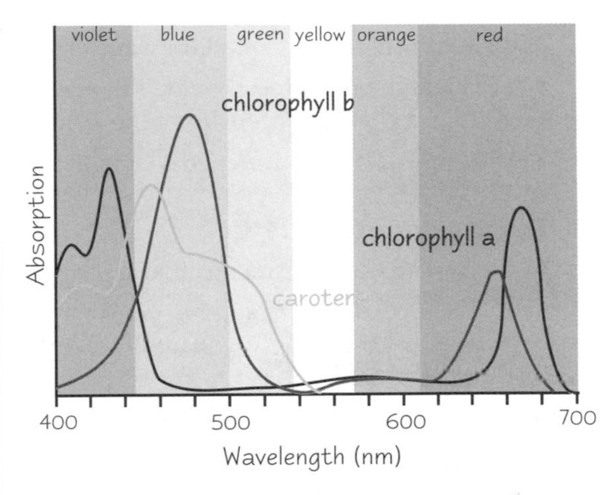

Having three different pigments allows the plant to absorb as many of the **different wavelengths** in white light as possible, as **efficiently** as possible. As the graph shows, the three pigments together are able to absorb most of the **red** and **blue** light in sunlight for photosynthesis. There aren't any pigments that absorb well in the **green** part of the spectrum though, so plants **reflect** green light instead of absorbing it. That's why they look green.

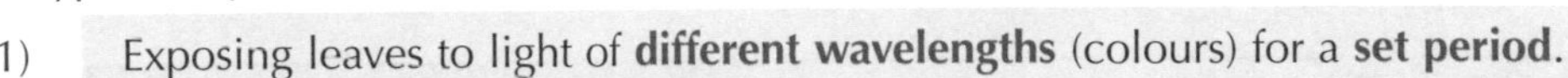

Action Spectra show the Rate of Photosynthesis at Different Wavelengths

Absorption spectra show which wavelengths are **absorbed** by plants. Further experiments have confirmed that these same wavelengths of light are then used in **photosynthesis** (pretty obvious I'd have thought, but hey).

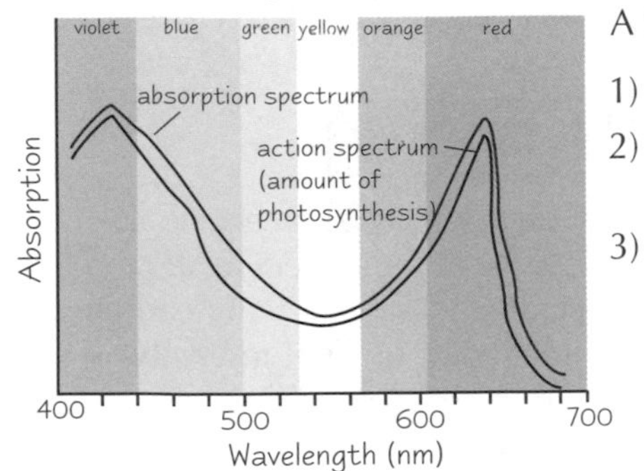

A typical experiment involves:

1) Exposing leaves to light of **different wavelengths** (colours) for a **set period**.
2) Measuring the **volume of oxygen** or **amount of carbohydrate** formed in each case (which corresponds to the amount of **photosynthesis** happening).
3) This is used to produce an **action spectrum** — a graph showing the **rate of photosynthesis** at **different wavelengths**.

There's a pretty **close correlation** between the **absorption** and **action spectra** for plants, as shown in the graph. They absorb mostly **blue** and **red** light, and these are also the wavelengths at which the most **photosynthesis** happens.

Practice Questions

Q1 What reaction occurs in the thylakoid membranes?

Q2 Name three pigments used in photosynthesis.

Q3 What does an absorption spectrum show?

Q4 Which pigment has a high absorption spectrum at about 680nm?

Q5 What does an action spectrum show?

Exam Questions

Q1 Describe how a leaf is adapted for photosynthesis. [5 marks]

Q2 Describe how a chloroplast is adapted for photosynthesis. [5 marks]

I'm sorry, chlorophyll, but we're just not on the same wavelength...

If you're not much of a physicist, don't start panicking just because you see the word 'wavelength' on this page. Light just travels in waves, and each colour of light has a different length of wave. But don't worry about any of that — just remember that plants have different pigments so they can absorb different colours and get as much light as possible for photosynthesis.

Factors Affecting Photosynthesis

Okay, at first glance this might seem quite a boring page — but I'm telling you now, appreciate it while it lasts...

There are **Optimum Conditions** for Photosynthesis

The **ideal conditions** for photosynthesis vary from one plant species to another, but the conditions below would be ideal for most plant species in temperate climates like the UK.

High light intensity of a certain **wavelength**

- Light is needed to provide the **energy** for the **light-dependent stage** of photosynthesis — the higher the **intensity** of the light, the **more** energy it provides.
- Only certain **wavelengths** of light are used for photosynthesis (see the **absorption spectrum** on page 73).

Temperature around **25°C**

- **Less** than **10 °C** means that enzymes are **inactive** — they need some energy from heat to function.
- **More** than **45 °C** and enzymes may start to **denature**. **Stomata** close at high temperatures to avoid losing too much water. This causes photosynthesis to slow because CO_2 can't enter the leaf when the stomata are closed.

Water — a constant supply is needed

- Too **little** and photosynthesis obviously has to stop, because it's one of the **reactants**.
- Too **much** waterlogs the soil and reduces uptake of minerals such as **magnesium**, which is needed to make **chlorophyll a**.

Carbon dioxide at 0.4%

- Carbon dioxide makes up **0.04%** of the gases in the atmosphere.
- Increasing this to **0.4%** gives a **higher rate** of photosynthesis, but any higher and the stomata start to **close**.

Light, **Temperature**, **Water** and CO_2 can all **Limit Photosynthesis**

All four of these things need to be at the right level to allow a plant to photosynthesise as quickly as it can. If any **one** of them is too low, it will be **limiting photosynthesis**. The other three factors could be at the perfect level, and it wouldn't make **any difference** to the speed of photosynthesis. On a warm, sunny, windless day, it's usually **carbon dioxide** that is the limiting factor in photosynthesis. At night, the limiting factor will obviously be **light intensity**. But **any** of those four factors could become the limiting factor, depending on the **environmental conditions**.

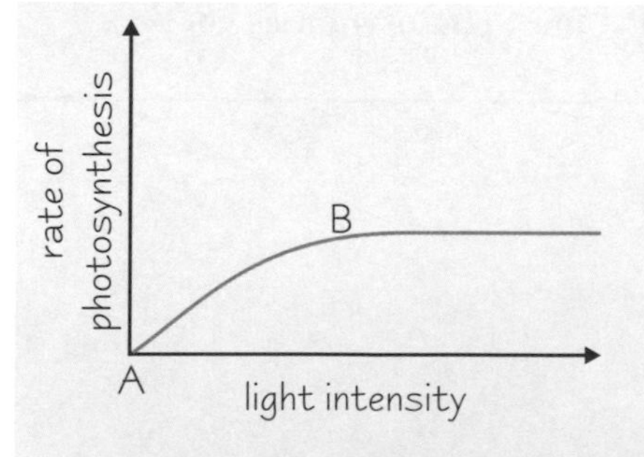

Between points A and B, the rate of photosynthesis is limited by the **light intensity**. So as the light intensity **increases**, so can the rate of photosynthesis. Point B is the **saturation point** — increasing light levels after this point makes no difference because **something else** has become the limiting factor. The graph now **levels off**.

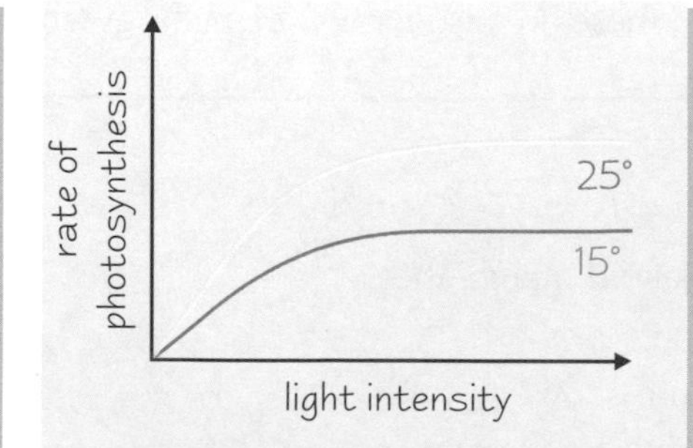

This graph shows that the rate of photosynthesis levels off when **light intensity** is no longer the limiting factor. The graph at **25 °C** levels off at a **higher point** than the one at **15 °C**, showing that **temperature** must have been a limiting factor at **15 °C**.

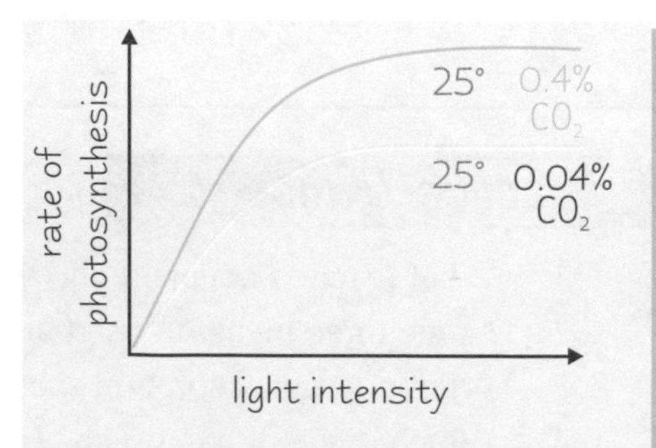

This graph shows that **carbon dioxide concentration** must have been a limiting factor when the previous graph levelled off at **25 °C**. This can be seen by comparing the two lines above, which are at the **same** light intensity and temperature.

Water is essential for photosynthesis and can also be a limiting factor, but lack of water causes lots of **other damage** to the plant — by the time it becomes a limiting factor in photosynthesis, that'll be the least of the plant's troubles.

Factors Affecting Photosynthesis

Plants can get CO_2 from Respiration

Plants **respire** as well as photosynthesise. Respiration **produces** carbon dioxide, and photosynthesis **uses it up**. Photosynthesis starts at dawn and the rate **increases** as the light gets **brighter**. At one point, **all** the carbon dioxide produced by respiration is **re-used** in photosynthesis, and this is known as the **compensation point**.

After this, **light** and **temperature** increase and respiration can't produce enough CO_2 for the amount of photosynthesis happening. The plant has to use CO_2 diffusing in from the **atmosphere** too. Then when dusk arrives and photosynthesis starts to slow down, the CO_2 from respiration is **all re-used** again. The **compensation point** has been reached again.

Plants get Water and Mineral Ions from soil

Remember from AS that plants absorb **water** and **dissolved mineral ions** from the soil, through their roots. This can happen by **diffusion**, but plants often **only** need **certain ions** and not others — diffusion **isn't** a selective process. So, ions needed by a plant are also absorbed by **active transport**. This means a plant can absorb more of a certain ion, even if it already has a **high concentration** of them **inside** its cells — so it doesn't have to rely on there being a **diffusion gradient** into the plant. It also means that the plant can **pump out** any ions it **doesn't need**.

Mineral ions travel up the plant with the water, in the **xylem**. Scientists know this because of experiments using **radioactive tracers**. These use radioactive forms of ions, so that the radioactivity can be detected and the scientists know where the ions have moved to.

Different Mineral Ions have different Functions in plants

Here are three of the most important mineral ions that plants absorb through their roots:

Mineral Ion	Function
nitrate	Needed to make amino acids for protein synthesis, and for organic bases for nucleic acids.
phosphate	Needed in photosynthesis and respiration reactions — for example there are phosphate groups in ATP and NADP. Also needed for phospholipids and in the backbones of DNA and RNA.
magnesium	Needed to make chlorophyll.

Practice Questions

Q1 Name three factors that can limit photosynthesis.

Q2 Name two sources of CO_2 for plants.

Q3 What is the compensation point?

Exam Questions

Q1 Explain whether a lack of nutrients in the soil can affect photosynthesis. [2 marks]

Q2 Would measuring the uptake of CO_2 by a plant give a true measurement of the rate of photosynthesis? Explain your answer. [2 marks]

What's stomata with you?

Stomata are really temperamental things... one minute they're too hot and they close, then they open, then the next minute there's too much CO_2 so they close again. Anyway I hope you appreciated this nice but dull page, because the next four pages are a wee bit trickier. Nothing to panic about, I hasten to add, but just... trickier. You'll see. Brace yourselves...

The Light-Dependent Reaction

Don't worry if this seems hard at first. Read it through carefully a couple of times, and it'll start to make sense.

Photosynthesis can be Split into Two Stages

Here's the overall equation for photosynthesis. Hopefully it'll look pretty familiar. When you were doing your GCSEs this little equation was all you had to worry about, but those days are long gone my friend.

$$6CO_2 + 6H_2O + \text{Energy} \xrightarrow{\text{chlorophyll}} C_6H_{12}O_6 + 6O_2$$

Photosynthesis happens in the **chloroplasts** (see page 72), and nowadays you need to know that it consists of two **stages**:

1) The **light-dependent reaction** (which, as the name suggests, needs **light energy**) takes place in the **thylakoid membranes** of the chloroplasts. Light energy is absorbed by pigments in the **photosystems** (see below), and used to provide the energy for the next stage — the light-independent reaction. There are **two different reactions** going on in this next stage — **cyclic photophosphorylation** and **non-cyclic photophosphorylation.** The plant can **switch between** the two, depending on whether it needs **reduced NADP** or just **ATP** (see below).

2) The **light-independent reaction** or **Calvin cycle** (which, as the name suggests, doesn't use light energy) happens in the **stroma** of the chloroplast. It does however use products from the light-dependent reactions — the **ATP** and the **reduced NADP** molecules produced supply the **energy** to make **glucose**. See pages 78-79 for more on the Calvin cycle.

The diagram shows how the two different reactions, light-dependent and light-independent, fit together in the chloroplast.

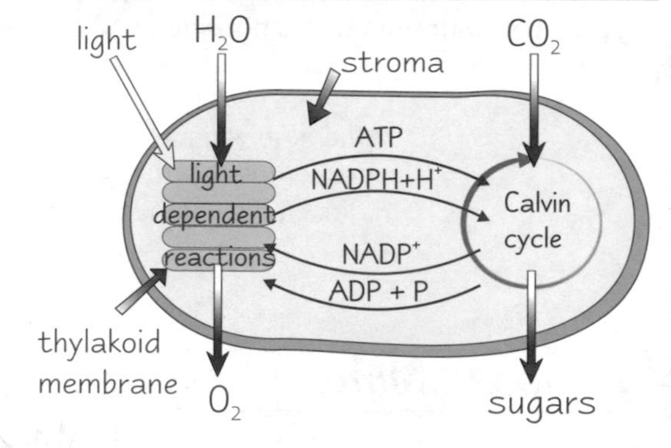

Photosystems I and II capture Light Energy

1) **Photosystems** are made up of **chlorophyll a**, **accessory pigments** (like **chlorophyll b** and **carotenoids**) and **proteins**. The proteins hold the pigment molecules in the best positions for **absorbing** light energy and **transferring** this energy to the **reaction centre** of the photosystem.
2) The **reaction centre** is a particular **chlorophyll molecule** called a **primary pigment**. The energy from absorbing light is passed from one **accessory pigment** to another until it reaches this **primary pigment**.
3) The energy is then used to **excite** pairs of **electrons** in the reaction centre pigment. The electrons move up to a higher **energy level**, ready to be used in the **light dependent reactions**.
4) There are **two** different photosystems used by plants to capture light energy. **Photosystem I** (or PSI) uses a chlorophyll molecule that absorbs light at wavelength **700 nm** in its reaction centre. PSI is found mostly in the **lamellae** in a chloroplast. **Photosystem II** (PSII) uses a chlorophyll molecule that absorbs light at around **680 nm** in its reaction centre. It's found mostly in the **thylakoids** of the chloroplast.

The Light-Dependent Reaction makes ATP in Photophosphorylation

The **energy** captured by the photosystems is used for **two** main things:

1) Making **ATP** from **ADP** and **inorganic phosphate** (**phosphorylation**). It's called **photo**phosphorylation here, as it uses **light**.
2) Splitting **water** into **H^+** ions and **OH^-** ions. This is called **photolysis**, because the splitting (lysis) is caused by light energy (photo). Photolysis is covered on the next page.

Photophosphorylation involves the **excited electrons** in the reaction centre of a photosystem being passed to a special molecule called an **electron acceptor**. These electrons are then passed along a **chain** of other electron carriers, each at a slightly **lower energy level** than the one before, so that the electrons **lose energy** at every stage in the chain. The energy given out is used to add a phosphate molecule to a molecule of ADP — and this is photophosphorylation.

You don't need to know the actual mechanism in detail, but it's very similar to how ATP is produced in **respiration**, involving the flow of **hydrogen ions** (H^+) through **stalked particles** (which you can read all about on page 7). The H^+ ions used in photosynthesis come from the **photolysis of water**, which is covered on the next page.

The difference between **cyclic** and **non-cyclic photophosphorylation** is in what happens to those electrons that have been moving through the chain of carriers. This is also explained on the next page.

The Light-Dependent Reaction

Cyclic Photophosphorylation just produces ATP

Cyclic photophosphorylation only uses **photosystem I**. It's called cyclic photophosphorylation because the electrons from the chlorophyll molecule are simply **passed back** to it after they've been through the chain of carriers — i.e. they're **recycled** and can be used repeatedly by the same molecule. This **doesn't** produce any **reduced NADP** (**NADPH + H⁺**) or O_2, but there **is** enough energy to make **ATP**. This can then be used in the **light-independent reaction**.

Non-cyclic Photophosphorylation produces ATP, NADPH and Oxygen

Non-cyclic photophosphorylation uses both **PSI** and **PSII**. It involves **photolysis**, which is the splitting of **water** using light energy. Photolysis only happens in **PSII**, because only PSII has the right **enzymes**.

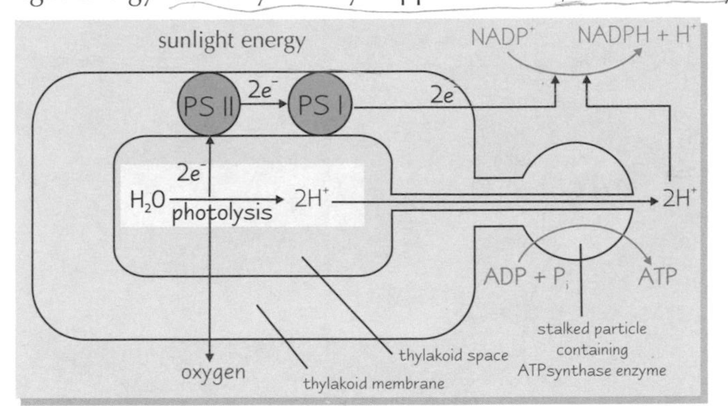

1) Light energy enters **PSII** and is used to move electrons to a **higher energy level**.
2) The electrons are passed along the chain of electron carriers to **photosystem I**. Most of the energy lost by the electrons during this process is used to make **ATP** (like in cyclic photophosphorylation).
3) Light energy is absorbed by PSI, which excites the electrons **again** to an **even higher energy level**.
4) The electrons are passed to a **different electron acceptor** and **don't** return to the chlorophyll.
5) For the chlorophyll to keep working, the electrons have to be replaced from **somewhere else** — so they're taken from a molecule of **water** (water is the electron donor). This makes the water molecule split up into **protons** (H^+) and **oxygen**.
6) The **protons** (H^+) from the water molecule combine with the **electrons** currently with the second electron acceptor to give **hydrogen atoms**. These are used to react with a substance called **NADP** to produce **NADPH** and **H⁺**. These are needed for the **light-independent reaction** (see pages 78-79).

So when electrons move back from the **second electron acceptor** to the chlorophyll molecule, that's **cyclic** photophosphorylation. If they don't, and the replacement electrons come from **water** instead, that's **non-cyclic**.

	cyclic photophosphorylation	non-cyclic photophosphorylation
photosystem	I	I and II
what's needed	light, ADP, inorganic phosphate	light, water, NADP, ADP, inorganic phosphate
what's produced	ATP	ATP, NADPH + H^+, O_2

Practice Questions

Q1 What is the full equation for photosynthesis?

Q2 Where in the chloroplast does the light-independent reaction of photosynthesis happen?

Q3 What is the reaction centre of a photosystem?

Q4 What two main things is the light energy captured by photosystems used for?

Q5 What is the difference between cyclic and non-cyclic photophosphorylation?

Q6 What useful waste product of photosynthesis is produced during non-cyclic photophosphorylation?

Exam Question

Q1 a) Where precisely in the plant does the light-dependent stage occur? [1 mark]

b) Which two compounds produced in the light-dependent stage are used in the light-independent stage? [2 marks]

c) Which of the light-dependent reactions of photosynthesis are involved in producing these compounds? [3 marks]

Photophosphorylate that, if you can...

*If you're feeling filled with despair as you read this tip, well, don't. You **will** understand this, don't give up. I guarantee it'll seem clearer every time you go through it, until at last you're left wondering what all the fuss was about. By the way, don't be put off when it says protons instead of hydrogen ions. That always confused me, but they mean the same thing.*

The Light-Independent Reaction

The second stage of photosynthesis doesn't need light energy, but that doesn't mean it can happily carry on in the dark. It still needs the NADPH + H⁺ and ATP from the light-dependent reaction, so if that stops then so does the light-independent stage.

The **Light-Independent** Reaction is also called the **Calvin Cycle**

The Calvin cycle makes **hexose sugars** (sugars with **6 carbons**, like **glucose** and **fructose**) from **carbon dioxide** and a **5-carbon** compound called **ribulose bisphosphate**. It happens in the **stroma** of the chloroplasts. There are a few steps in the reaction, and it needs **energy** and **H^+ ions** to keep the cycle going. These are provided by the products of the **light-dependent reaction**, **ATP** and **reduced NADP** (**NADPH + H^+**).

The diagram shows what happens at each stage in the cycle. The numbers in brackets (5C, 3C etc.) show how many **carbon atoms** there are in each molecule — the cleverest bit of the cycle is how it turns a **5-carbon** compound into a **6-carbon** one.

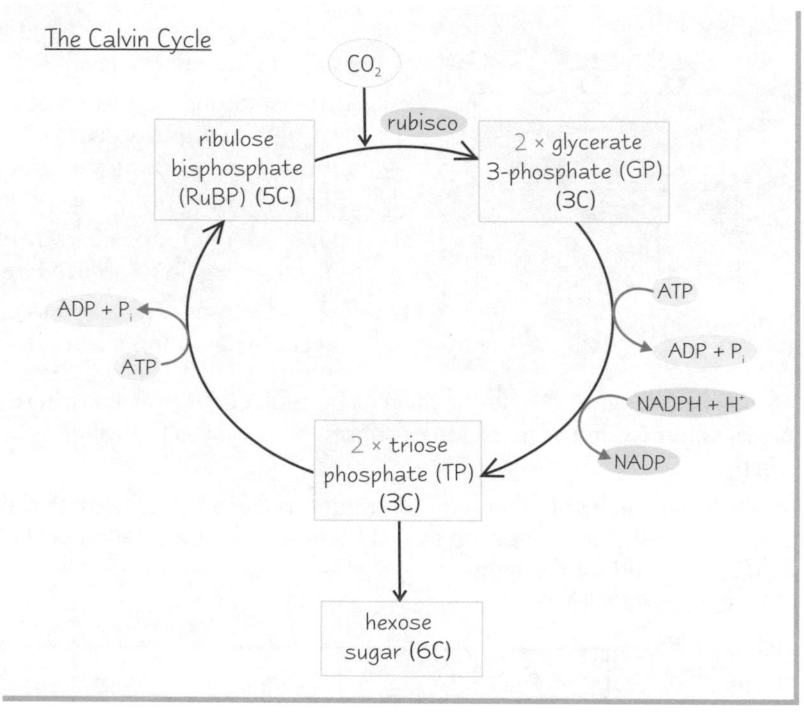

1) **CO_2** enters the leaf through the **stomata** and diffuses into the **stroma** of the chloroplast.
2) There it's taken up by **ribulose bisphosphate** (**RuBP**), a **5-carbon** compound. This gives an **unstable 6-carbon** compound, which quickly breaks down into **two** molecules of a **3-carbon** compound called **glycerate 3-phosphate** (**GP**).
3) This reaction is catalysed by the enzyme **ribulose bisphosphate carboxylase** (**rubisco**).
4) **ATP** from the **light-dependent stage** of photosynthesis is now used to provide the energy to turn the **3-carbon** compound, **GP**, into a **different** 3-carbon compound called **triose phosphate**.
5) This reaction also needs **H^+ ions**, which are provided by the **reduced NADP** (**NADPH + H^+**) made in the **light-dependent reaction**.
6) **Two** triose phosphate molecules then **join together** to give **one hexose sugar** (e.g. glucose).

Five out of every **six** molecules of **triose phosphate** produced in the Calvin cycle are **not** used to make hexose sugars, but to **regenerate** RuBP. Making RuBP from triose phosphate molecules uses the rest of the ATP produced by the light-dependent reaction.

The Calvin cycle is the starting point for making **all** the substances a plant needs — plants can't take in **proteins** and **lipids** like animals can.

- Other **carbohydrates**, like **starch**, **sucrose** and **cellulose**, can be easily made by joining the simple hexose sugars together in different ways.
- **Lipids** are made using **glycerol** synthesised from **triose phosphate**, and **fatty acids** from **glycerate 3-phosphate**.
- **Proteins** are made up of **amino acids**, which are also synthesised from **glycerate 3-phosphate**.

The Light-Independent Reaction

The Calvin Cycle needs to turn 6 times to make 1 Glucose Molecule

On the last page, you saw that **five** out of every **six triose phosphate molecules** go back into the cycle to regenerate **RuBP**, rather than going to make new hexose sugars. This means that the cycle has to happen **six times** just to make one new sugar. The box below shows why.

Remember, the photosynthesis equation uses 6 CO_2.

1) **6 RuBP** and **6 CO_2** molecules (i.e. 6 turns of the cycle) convert a total of **12 glycerate 3-phosphate** molecules into **12 triose phosphate molecules**.
2) **Two** molecules of **triose phosphate** are removed from the cycle and make **one glucose molecule**.
3) **10 triose phosphate** molecules regenerate the RuBP.

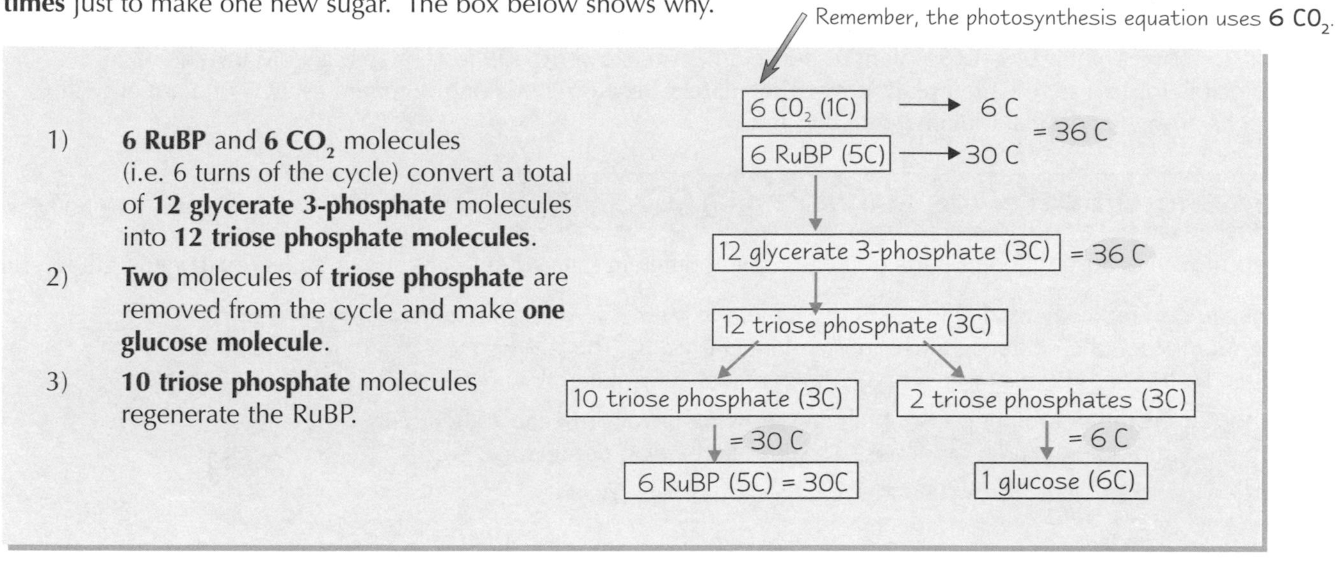

This might seem a bit inefficient, but it keeps the cycle going and makes sure that there's always **enough RuBP** there ready to combine with CO_2 taken in from the atmosphere.

Practice Questions

Q1 Give two examples of a hexose sugar.

Q2 How many carbon atoms are there in a molecule of triose phosphate?

Q3 Name the enzyme that catalyses the reaction between carbon dioxide and ribulose bisphosphate.

Q4 How is the Calvin cycle involved in making lipids?

Q5 How many CO_2 molecules need to enter the Calvin cycle to make one molecule of glucose?

Exam Questions

Q1 Which molecule in photosynthesis:
a) is the carbon dioxide acceptor? [1 mark]
b) provides the hydrogen ions to reduce glycerate 3-phosphate? [1 mark]
c) is regenerated in the Calvin cycle? [1 mark]
d) is known as rubisco? [1 mark]
e) is made of 5 carbon atoms? [1 mark]

Q2 Look at the diagram on the right and describe what is happening:
a) between points a and b, [1 mark]
b) between points b and c, [1 mark]
c) at point c. [1 mark]

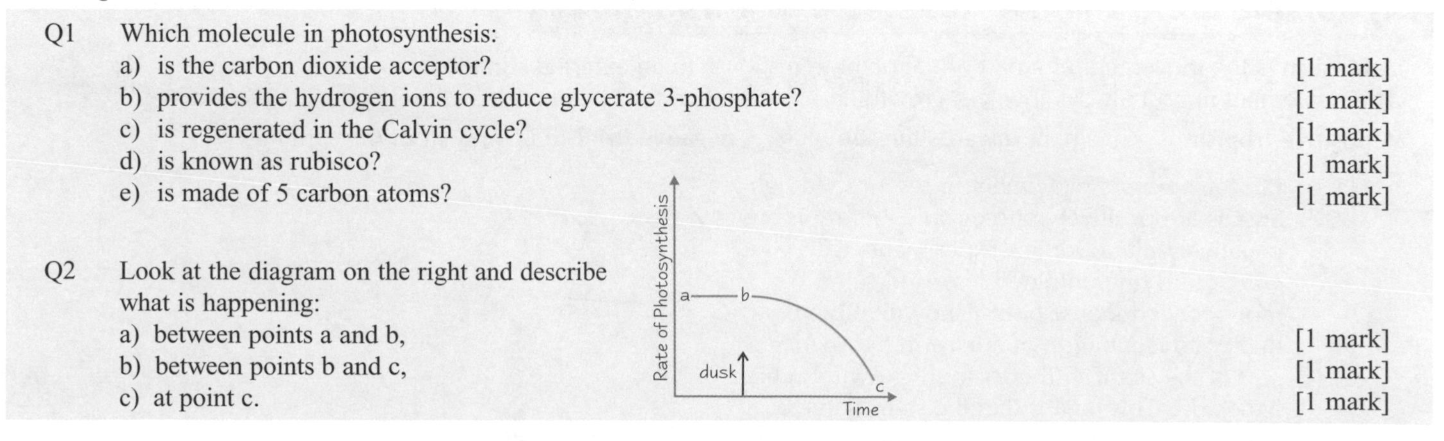

Don't worry — I promise the next section will be a lot easier...

Now don't start sinking into despair again. I know it's a lot to take coming after that last page, but actually this one is probably a bit easier. Learn that cycle on the last page and you're three quarters of the way there. Don't worry too much about learning the maths bit in the box on this page either — as long as you've got the general gist, that's enough.

Control of Growth in Plants

Plants don't have brains, but they do respond to their environment. They move too, but they're so slow you can't see them do it. Just as well because plants that ran around the place would be pretty scary things.

Plants respond more Slowly than Animals

Coordination and **response** in plants happens by **chemical communication** between cells. It takes time for the chemicals to move to their targets, so plant response times are **slower** than those of animals. Many chemical communicators in plants are called **plant growth regulators** because they control certain aspects of plant growth, making cells grow bigger or stimulating cell division.

Auxins and Gibberellins stimulate Cell Growth

Plants can move because different parts of them grow at different rates, due to chemicals called **auxins** and **gibberellins**.

Auxins are continuously made at the **shoot tips** and in **young leaves**. They diffuse from cell to cell and loosen the cellulose fibres in the plant cell walls. This allows more water to be taken up by the cells by **osmosis**. The cell then makes new material, and **elongates**.

The shoot tip is the dominant part of the plant growing upwards because the auxins inhibit growth of lateral (side) branches. This is called **apical dominance**.

A **drop** in auxin levels is also what causes leaves to fall (**abscission**).

Gibberellins are also produced in parts of the plant associated with growth, like **young leaves** and the **embryos** in seeds. They stimulate the synthesis of **amylase**, which **hydrolyses** stored starch to **maltose** for respiration. Many dormant seeds secrete gibberellins when they are soaked in water, which encourages them to germinate. Gibberellins also allow stem elongation, which enables plants to grow tall.

Auxins and gibberellins are synergistic — they work together to produce the same effect. Antagonistic substances oppose each other's actions.

Auxins are used **commercially** by farmers and plant breeders for the following uses:

1) **Selective weedkillers.** Auxins can kill fast-growing weeds. This is because they make the weed use up its stored chemical energy to produce a long stem, instead of leaves for photosynthesis.
2) **Rooting hormones.** Specific kinds of auxins promote the rooting of cuttings.
3) Applying low concentration of auxin in the early stages of **fruit production** prevents the fruit from falling. Applying a high concentration at a later stage promotes fruit drop.

Tropisms are a Plant's Response to a Stimulus

A **tropism** is the movement of a part of the plant in response to an **external stimulus**.
The movement almost always involves growth and so it's influenced by plant growth regulators.

A **positive tropism** is movement **towards** the stimulus. A **negative tropism** is **away** from the stimulus:

1) **Phototropism** is movement in response to **light**. Shoots are positively phototropic, but roots are negatively phototropic. Experiments with **coleoptiles** (the unfolded leaves of monocotyledonous shoots) show that there's a higher concentration of auxins in the shaded part of the stem, so there's more growth on the dark side. This means that the stem starts to curve towards the light.

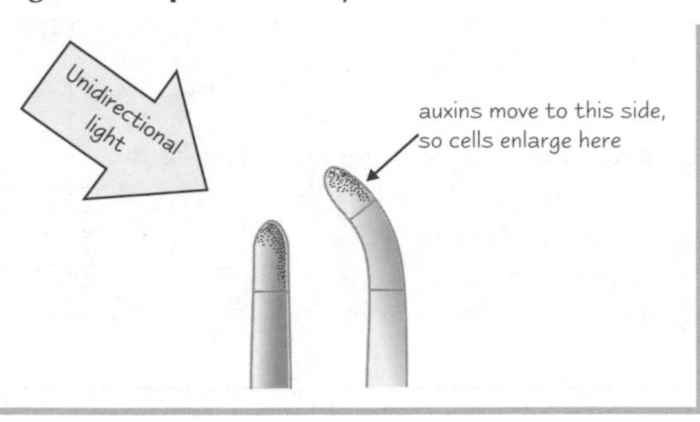

2) **Geotropism** is movement due to **gravity**. Shoots are **negatively geotropic** (they grow upwards) but roots are **positively geotropic** and grow downwards.

Control of Growth in Plants

Abscisic Acid and *Ethene* are *Antagonistic* to *Auxins and Gibberellins*

1) **Abscisic acid** is produced in most parts of the plant and prepares the plant for dormancy (the period when they don't grow) by **inhibiting growth**. It stimulates the **closure** of the **stomata** in leaves when water is in short supply, and **inhibits germination** in seeds.
2) **Ethene** is also produced in most plant parts — it breaks bud dormancy and stimulates the **ripening** of fruit.

Cytokinins stimulate cell division

Chemicals called **cytokinins** stimulate **mitosis** by increasing DNA and RNA synthesis. They are produced in **seeds** to encourage embryo growth, but are also made in the roots and carried to other parts of the plant.

Phytochromes control *Flowering Times*

When **light** intensity or wavelength determines the timing of a biological process, this is called **photoperiodism**. The flowering time of many plants is determined by the length of the day (the **photoperiod**) and controlled by chemicals called **phytochromes**.

Different species react in different ways to photoperiods:
1) **Long-day plants** flower in the summer (long days, short nights).
2) **Short-day plants** flower in the winter (short days, long nights).
3) **Day-neutral plants** flower independently of day length.

Phytochrome exists in two forms:
1) **Phytochrome red (PR)** absorbs **red light** (660nm wavelength); and changes into the **PFR form**.
2) **Phytochrome far red (PFR)** absorbs **far red light** (730nm wavelength) and changes slowly back into **PR** in the dark.

Example — long-day plants

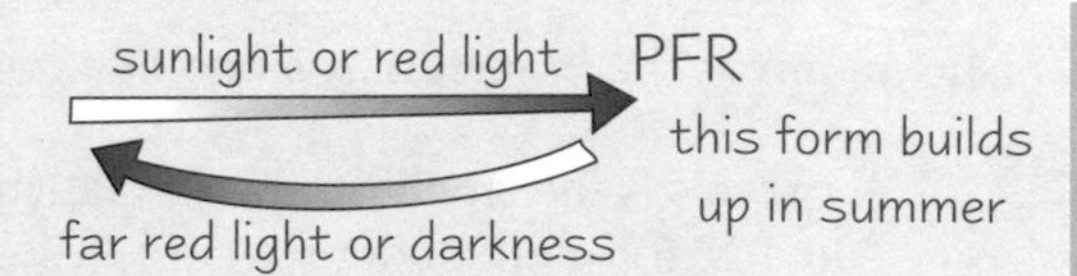

Sunlight contains more **red** than **far red** wavelengths, so at the end of the longer **summer** days a lot of PR has been changed into PFR. Because the nights are **short**, there's not enough time for PFR to be changed back into PR, so there's more **PFR**, which **stimulates flowering in long-day plants**.
In winter (when days are **short**), the nights are long enough for a lot of PFR to be converted back to PR. So there's a **lack** of PFR (which inhibits flowering in short-day plants).

In short-day plants, PFR may inhibit flowering, so they flower in winter.

Practice Questions

Q1 Distinguish between the effects of auxin and gibberellin on plant cell growth.
Q2 Auxins control apical dominance. Explain what this means.
Q3 Explain how the distribution of auxin in a growing shoot enables it to grow towards a light source.
Q4 Give one commercial application of auxins.

Exam Questions

Q1 Explain how the production of gibberellins in seeds stimulates them to germinate. [5 marks]
Q2 With reference to phytochromes, suggest how increased day length stimulates the flowering of some species of plants in the summer. [5 marks]

It's all gibberellin-ish to me...

Phytochromes are probably the most complicated things on these pages. The most confusing thing about it is that not all plants respond to phytochromes in the same way. If you concentrate on how long-day plants react to phytochromes, and get that part clear in your head first, then you should be well on your way to understanding the pesky little chemicals.

Classification

Classification is all about grouping together organisms that have similar characteristics.
The system of classification in use today was invented by a Swedish botanist, Carolus Linnaeus, in the 1700s.

Classification is the way Living Organisms are Divided into **Groups**

The classification system in use today puts organisms into one of five **kingdoms**:

KINGDOM	EXAMPLES	FEATURES
Prokaryotae	bacteria	unicellular, no nucleus, less than 5 μm, naked DNA in circular strands, cell walls of peptidoglycan
Protoctista	algae, protozoa	eukaryotic cells, usually live in water, unicellular or simple multicellular
Fungi	moulds, yeasts and mushrooms	eukaryotic, heterotrophic, chitin cell wall, saprotrophic
Plantae	mosses, ferns, flowering plants	eukaryotic, multicellular, cell walls made of cellulose, photosynthetic, contain chlorophyll, autotrophic
Animalia	nematodes (roundworms), molluscs, insects, fish, reptiles, birds, mammals	eukaryotic, multicellular, no cell walls, heterotrophic

You can **Classify Organisms** according to how they **Feed**

There are **three** main ways of getting **nutrition** —

1) **Saprotrophic** organisms, e.g. **fungi**, absorb substances from **dead** or **decaying** organisms using **enzymes**.
2) **Autotrophic** organisms, e.g. **plants**, produce their **own** food using **photosynthesis**.
3) **Heterotrophic** organisms, e.g. **animals**, consume complex organic molecules, i.e they consume **plants** and **animals**.

All Organisms can be organised into **Taxonomic Groups**

Taxonomy is the branch of science that deals with **classification**.

A **species** is the **smallest** unit of classification (see pages 106-107 for more about species). Closely related species are grouped into **genera** (singular = genus) and closely related genera are grouped into **families**. The system continues like this in a hierarchical pattern until you get to the largest unit of classification, the **kingdom**.

The Hierarchy of Classification
Kingdom
Phylum
Class
Order
Family
Genus
Species

For Example, Humans are **Homo sapiens**

This is how **humans** are classified:

		FEATURES
KINGDOM	Animalia	animal
PHYLUM	Chordata	has nerve cord
CLASS	Mammalia	has mammary glands and feeds young on milk, has hair / fur
ORDER	Primates	finger and toe nails, opposable thumb, reduced snout and flattened face, binocular vision, forward-facing eyes
FAMILY	Hominidae	relatively large brain, no tail, skeleton adapted for upright or semi-upright stance
GENUS	*Homo*	cranial capacity > 750 cm^3, upright posture
SPECIES	*sapiens*	erect body carriage, highly developed brain, capacity for abstract reasoning and speech

This column shows the features that have been used to classify humans into each of these groups.

Classification

The **Binomial System** is used to **Name** organisms

The full name of a human is **Animalia Chordata Mammalia Primate Hominidae *Homo sapiens***. The name gives you a lot of information about how humans have been classified. Using full names is a bit of a mouthful so it's common practice to just give the **genus** and **species** names — that's the **binomial** ('two names') **system**.

The binomial system has a couple of **conventions**:

1) Names are always written in ***italics*** (or they're **underlined** if they're **handwritten**).
2) The **genus** name is always **capitalised** and the **species** name always starts with a **lower case** letter.

e.g.

Human	*Homo sapiens*
Polar bear	*Ursus maritimus*
Sweet pea	*Lathyrus odoratus*

Cladograms show **Evolutionary Relationships**

When taxonomy was first developed organisms were classified according to characteristics that were **easy to observe**, for example, number of legs. Thanks to modern **scientific techniques** like **DNA technology**, **genetics**, **biochemical analysis** and **behavioural analysis**, many more **criteria** can now be used to classify organisms.

A **cladogram** is a diagram that emphasises **phylogeny** (the genetic relationship between organisms). Cladistics focuses on the features of organisms that are **evolutionary developments**. The **advantage** of cladograms is that you can see points where **one** species split into **two**.

A CLADOGRAM OF PRIMATE EVOLUTION

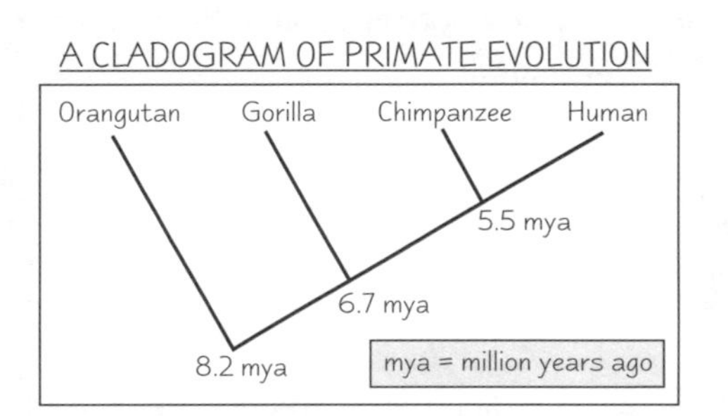

99.4% of a chimp's DNA is identical to human DNA. The fact that humans and chimps diverged most recently shows that humans are more closely related to chimps than they are to the other ape species in the cladogram.

Practice Questions

Q1 Name the five kingdoms of classification, giving an example organism in each.

Q2 What do the phrases saprotrophic, autotrophic and heterotrophic mean?

Q3 Explain the difference between fungi and plants in terms of how they get their nutrition.

Q4 What are the two rules for using the binomial system?

Exam Questions

Q1 Explain the difference between phylogenetic classification and traditional classification. [2 marks]

Q2 The King Penguin has the scientific name *Aptenodytes patagonicus*. Fill out the missing words a) – e) in the table. [5 marks]

Kingdom	Animalia
a)	Chordata
b)	Aves
Order	Sphenisciformes
c)	Spheniscidae
Genus	d)
Species	e)

I prefer scantilycladograms...

The good thing about this is everything is pretty straightforward — don't be put off if lots of the words are new to you (and if 'cladogram' is part of your day-to-day vocabulary then I suggest you get out more). You need to learn this thoroughly. In the exam you'll be glad that you did cos there's often some easy marks to be had about this kind of stuff.

Investigating Ecosystems

If you've been on an ecology field trip you'll be familiar with this stuff. You'll be relieved to know that you can revise this in the comfort of your own bedroom — you won't be asked to stand in a river catching horrible squirmy things.

You need to know how to take **Abiotic Measurements**

Abiotic factors (non-living factors) affect the distribution of organisms in the environment (see page 88 for more). So you need to be able to measure them in the field to investigate these distributions.

Temperature is easy enough — just use a **thermometer**.

pH measurements are only taken for soil or water. **Indicator** paper or an electronic **pH monitor** are used.

Light intensity is difficult to measure because it varies a lot over short periods of time. You get the most accurate results if you connect a **light sensor** to a data logger and take readings over a period of time.

Oxygen level only needs to be measured in aquatic habitats. An **oxygen electrode** is used to take readings.

Air humidity is measured with a hygrometer.

Moisture content of soil is calculated by finding the mass of a soil sample and putting it in an oven to dry out. The amount of mass that has been lost is worked out as a % of the original mass.

Quadrat Frames are a Basic Tool for Ecological Sampling

1) To investigate the species found in a particular area, you can use a piece of equipment called a **quadrat frame** — a square frame made from metal or wood. The area inside this square is known as a **quadrat**.
2) **Quadrat frames** are **laid on the ground** (or the river / sea / pond bed if it's an aquatic environment). The **total number** of **species** in the quadrat frame is recorded as well as the number of **individuals** of each species.
3) Generally it's not practical to collect data for a whole area (it would take you ages) so **samples** are taken instead. This involves measuring lots of quadrats from different parts of the area. The data from the samples is then used to **calculate** the figures for the **entire area** being studied. **Random sampling** is used to make sure that there isn't any **bias** in the data (this involves dividing the study area into a grid and then using a random number generator to select the coordinates at which quadrat frames are placed).

It's also important to consider the **size** of the quadrat — smaller quadrats give more accurate results, but it takes longer to collect the data and they're not appropriate for large plants and trees.

Point Quadrats are an alternative to quadrat frames

Pins are dropped through holes in the frame and every plant that each pin hits is recorded. If a pin hits several **overlapping** plants, **all** of them are recorded. A tape measure is laid along the area you want to study and the quadrat is placed at regular intervals (e.g. every 2 metres) at a right angle to the tape.

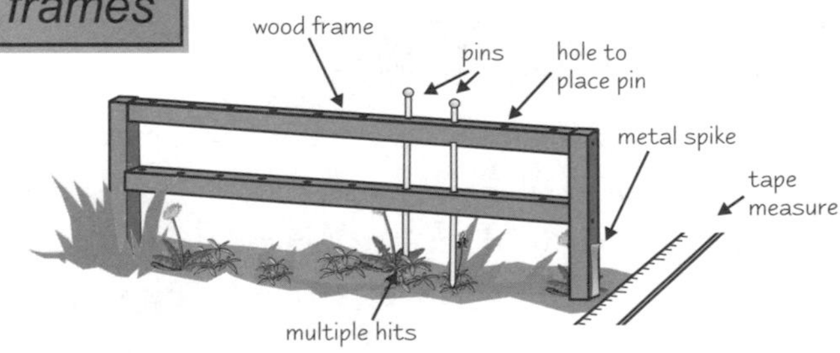

Line and **Belt Transects** are used to **Survey** an area

The line you select to sample across the area is called a **transect**. Transects are useful when you want to look for **trends** in an area e.g. the **distribution of species** from low tide to the top of a rocky shore.

A **line transect** is when you place a tape measure along the transect and record what species are touching the tape measure.

A **belt transect** is when data is collected between two transects a short distance apart. This is done by placing frame quadrats next to each other along the transect.

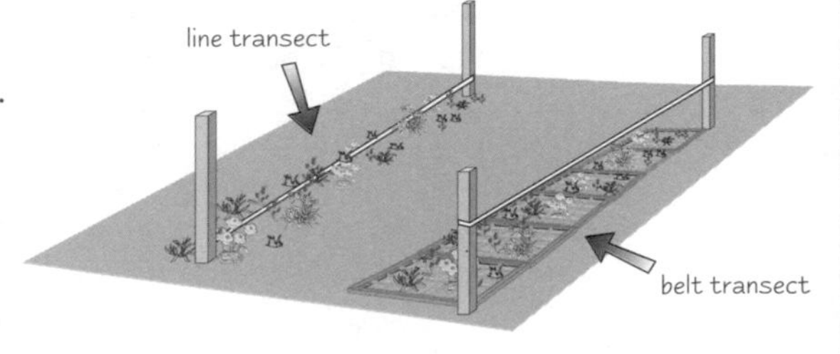

If it would take ages to count all the species along the transect, you can take measurements at set intervals, e.g. 1 m apart. This is called an **interrupted transect**.

The data collected from belt or line transects is plotted on a **kite diagram** (that's just a fancy kind of graph) and trends across the area can be observed.

Investigating Ecosystems

To sample **Animals**, you've got to **Catch** them

Most animals are **mobile** so they can't be sampled using quadrats or transects. There's various methods for catching animals depending on their **size** and the **kind of habitat** being investigated.

Nets can be used to trap flying insects and aquatic animals.

Pitfall traps can be used to catch walking insects on land. The insects fall into the trap and are... well, trapped.

Pooters are used to catch individual insects which are chosen by the user.

Tullgren funnels are used to extract small animals from soil samples. The animals move away from the light and heat produced by the bulb and eventually fall through the barrier into the alcohol below the funnel.

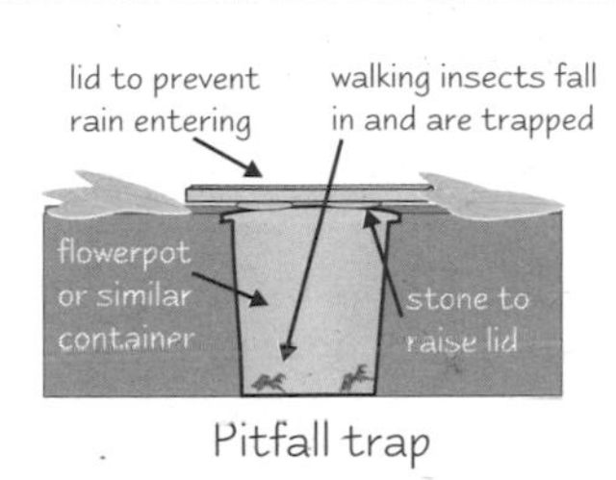

Pitfall trap

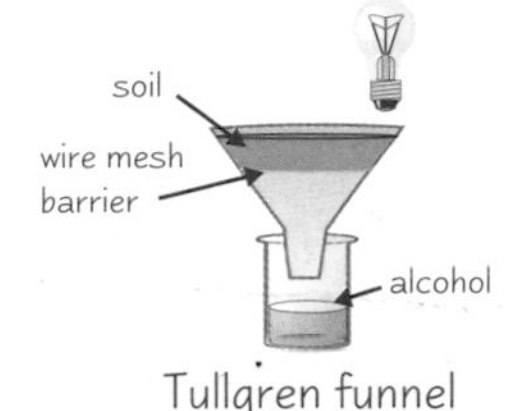

Tullgren funnel

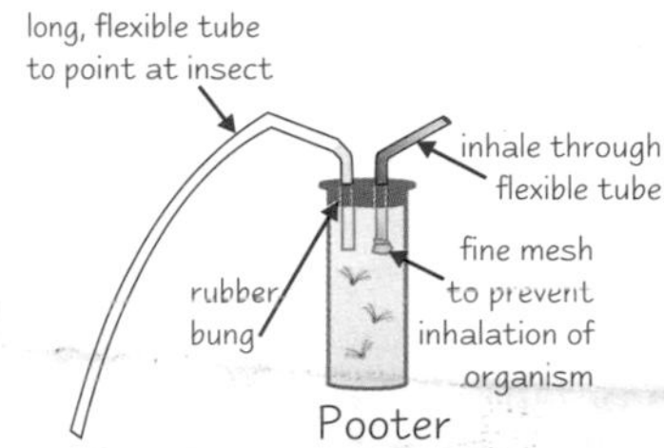

Pooter

The **Mark-Release-Recapture** Technique is for estimating **Population Sizes**

The mark-release-recapture method is basically this:

1) **Capture** a sample of the population.
2) **Mark** them in a harmless way.
3) **Release** them back into their habitat.
4) Take a **second sample** from the population.
5) **Count** how many of the second sample are marked.
6) Estimate the **size** of the whole population using the **Lincoln index**.

$$\text{Population size (S)} = \frac{n_1 \times n_2}{n_m}$$

n_1 = number of individuals in first sample
n_2 = number of individuals in second sample
n_m = number of marked individuals in the second sample

The accuracy of this method depends upon these **assumptions**:

1) The marked sample has had enough **time** and **opportunity** to **mix** back with the population.
2) The marking has not affected the individuals' **chances of survival**.
3) **Changes** in population size due to **births**, **deaths** and **migration** are **small**.
4) The marking has **remained visible** in all cases — so it needs to be waterproof.

Good ways of marking animals include using a UV pen or cutting a little bit of the animal's fur off.

Practice Questions

Q1 How would you calculate the moisture content of soil?

Q2 What is the difference between a line transect and a belt transect?

Q3 What piece of apparatus is used to extract small organisms from a soil sample?

Exam Questions

Q1 When measuring light intensity in an ecosystem, why is it not sufficient to take a single light-meter reading? [3 marks]

Q2 Under what circumstances would you use a transect rather than random sampling of an ecosystem? Give an example in your answer. [2 marks]

Q3 A population of woodlice was sampled using pitfall traps. 80 individuals were caught. The sample was marked and released. Three days later, a second sample was taken and 100 individuals were captured. Of these, 10 had marks. Use the Lincoln index to estimate the size of the woodlouse population. [2 marks]

What do you collect in a poo-ter again?

Aren't you glad that we don't use the mark-release-recapture technique to measure our population size. I don't fancy falling in a pitfall trap and then getting a chunk of my hair cut off. Seems kind of barbaric, now I think about it. When we did this experiment at school, we never caught any of the woodlice we'd marked again. They'd all disappeared...

Succession

Succession is all about how ecosystems change over time. Apart from a few fancy words I think that it's one of the easiest things in A2 biology — it's a lot more straightforward than the Krebs cycle and photosynthesis.

*There are **Two** different types of **Succession***

Succession is sometimes called ecological succession in exams.

Succession is the process where **plant communities** gradually develop on **bare land**.
Eventually a **stable climax community** develops and after that big changes don't tend to happen.
There are **two** different types of succession...

Primary succession

Happens on land where there is no proper soil and **no living organisms**. New land created by a **volcanic eruption** is a good example of a place where primary succession will occur.

Secondary succession

Happens when most of the living organisms in an area are **destroyed** but the **soil** and **some** living organisms remain. Examples include: woodland that has been burned by a **forest fire**, areas subject to severe **pollution** or land that is cleared by **people** for things like **housing** or **new roads**.

Each stage in the succession of an area is called a **seral stage**. In every seral stage the plants change the environmental conditions, making them suitable for the next plants to move in.

Sand Dunes** to **Woodland** is an example of **Primary Succession

This example shows the seral stages that change **bare** sand dunes into **mature woodland**.

The first plants to colonise an area have to be specialised so they can deal with the harsh abiotic conditions. These plants are known as pioneer species. They are usually herbaceous (non-woody).

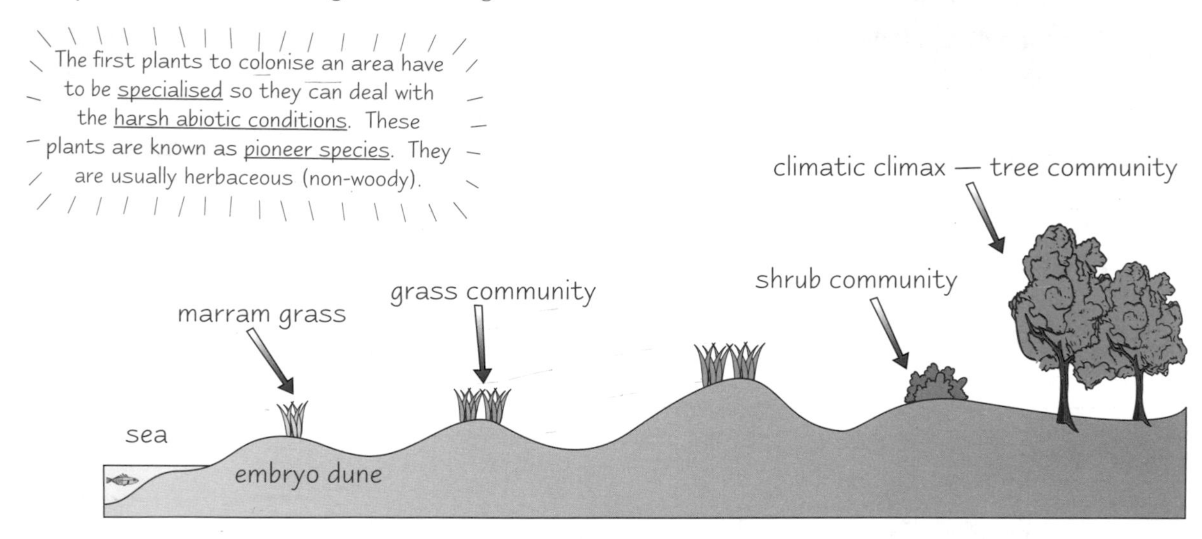

1) The first **'pioneer'** species to colonise the area need to be able to cope with the **harsh abiotic conditions** on the **sand dunes** — there is **little fresh water** available, there are **high salt levels**, the **winds** are **strong** and there is **no proper soil**. Marram grass is well adapted to dry conditions so it is usually the first to start growing.
2) As the pioneer species begin to **die** they are broken down by microorganisms. The dead marram grass adds **organic material** to the sand creating a very basic **'soil'** which can hold more water than plain sand.
3) This soil means that the abiotic conditions are **less hostile** and so other, less specialised grasses begin to grow.
4) These new grasses will eventually **out-compete** the original colonisers via **interspecific** competition (see p. 88).
5) As each new species moves in, more **niches** (see p. 88) are created making the area suitable for even more species.
6) After the grass communities have all been out-competed, the area will be colonised by **shrubs** like **brambles**.
7) Eventually the area becomes dominated by **trees** — in Europe the trees will usually be things like **birch** and **oak**. The trees dominate because they prevent light from reaching the herbaceous plants below the leaf canopy.

Succession

Diversity Increases and Species Change as succession progresses

When succession happens in any environment the general pattern of change is always the same:

- The species present become **more complex** e.g. a forest starts with simple mosses and finishes with trees.
- The **total number** of organisms **increases**.
- The **number of species increases**.
- **Larger species** of plants arrive.
- **Animals** begin to move into the area — with each seral stage **larger** animals move in.
- **Food webs** become more **complex**.
- Overall, these changes mean that the **ecosystem** (see p. 88) becomes more **stable**.

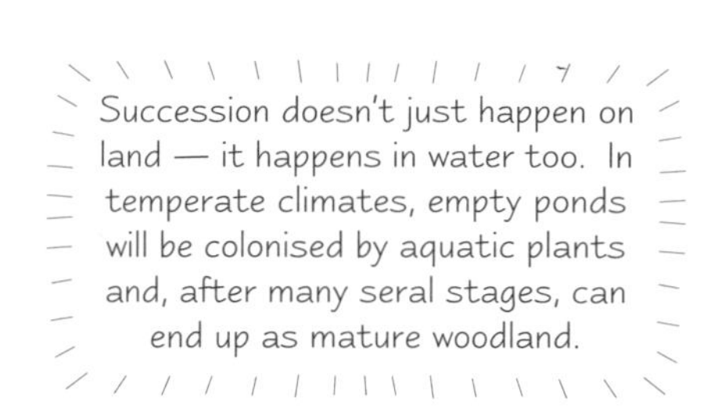

There are different Types of Climax Community

Various factors can **stop succession** going any further and lead to a **climax** community.
The climax is **classified** according to **what** has prevented the succession from going any further...

1) In a **climatic climax**, the succession has gone as far as the **climate** in the area will allow.
E.g. Trees can't grow at high altitudes, so high up on alpine mountains the largest plants are **shrubs**.
2) **Human activities** can stop succession by felling trees, ploughing fields or grazing animals on farmland.
Some ecosystems are deliberately '**managed**' to keep them in a particular state, for example, the heather on moorland is burned every 5-7 years to prevent woodland from developing. When succession is stopped **artificially** like this the climax community is called a **plagioclimax**.

Farmland is an example of a Plagioclimax

Succession is stopped by regular **ploughing** or by the **grazing** of stock. In a grazed field, **grass** can survive because it is fast-growing, but slow growing plants get eaten before they can get established.

If the **grazing stops**, then slower growing plants can gradually begin to establish themselves. As they do, the **grass** will become **less dominant**. The new plant species will attract a wider range of **insects** and so the area will **increase** its **diversity**.

Eventually, the field will be replaced by **woodland** by secondary succession. The process of succession is exactly the same for primary and secondary succession, they just have different starting points.

Practice Questions

Q1 What is secondary succession?

Q2 What are the stages in succession called?

Q3 What is a pioneer species?

Q4 What name is given to a climax community brought about by human intervention?

Exam Questions

Q1 Define the term 'ecological succession', explaining how it occurs and the different types of climax communities that can be produced. [8 marks]

Q2 a) Suggest three features of a plant species that might make it a successful pioneer species. [3 marks]
b) Suggest two reasons why such a pioneer species may disappear during the early stages of succession. [2 marks]

There are many different types of climax...

If your enthusiasm for all these Biology facts is waning, why not try reading some ICT... "Remote management supports users on a network. In the event of a significant problem or recurring error, the network administrator takes temporary control. Afterwards, the user can be advised on how to avoid similar problems in the future..." You see what I'm saying?

Populations

Some terms to know and some graphs to interpret, but these two pages are more common sense than rocket science.

You need to learn some **Definitions** to get you started

Ecosystem An **ecological unit** which includes all the **organisms** living in a particular area and all the **abiotic** (non-living) features of the local environment.

Population A group of individuals of the same species living in the same area.

- **Population size** — The total number of individuals in a population.
- **Population density** — The number of individuals per unit of the area inhabited by the population.

Community All the **living organisms** in an ecosystem. These organisms are all **interconnected** by food chains and food webs.

Environment The **conditions** surrounding an organism, including both abiotic factors (e.g. temperature, rainfall) and biotic factors (e.g. predation, competition).

Niche The **'role'** an organism has in its environment — where it lives, what it eats, where and when it feeds, when it is active etc. Every species has its own **unique** niche.

The opposite of abiotic is biotic (to do with living things).

Population Size is affected by **Limiting Factors**

Limiting factors are things that put an **upper limit** on a population's size.

Many limiting factors are **abiotic** (non-living) — examples include: **temperature**, **soil** or **water pH**, **oxygen availability**, and the supply of **mineral nutrients**.

Some of these things (e.g. temperature and oxygen availability) can affect the **rate** of **population growth** as well as limiting the total **size**.

The maximum population size that an environment can support is called the **carrying capacity**.

Check exam questions about limiting factors to see if you're being asked about growth rates or maximum sizes.

Limiting factors can be classified as either **density-dependent** or **density-independent**:

Density-dependent factors

- The effects of these factors depend on the **population density**.
- They have a greater impact as a population gets more **dense**.
- **Competition**, **predation** and **disease** are examples of **density-dependent limiting factors**.
- These factors are usually **biotic**.

Density-independent factors

- These factors are completely **unrelated** to **population density**. The effect of the factor will be the same whether the population density is low or high.
- **Flooding**, **forest fires** and **temperature** are examples of **density-independent limiting factors**.
- These factors are usually **abiotic**.

Environmental resistance is the **prevention** of **unrestricted growth** of a population by the combined action of **abiotic** and **biotic factors** (i.e. it's when limiting factors stop the population growing and growing).

Competition can be **Interspecific** or **Intraspecific**

1) Competition occurs when a lot of organisms are competing for some sort of limited **'resource'** — very often it's food, but it can be other things like **shelter**, **nesting sites** and **mates**.
2) If the organisms competing are of the **same** species, it is called **intraspecific** competition. If they're from **different** species, it's called **interspecific** competition.

Populations

Interspecific Competition affects population Distribution

Sometimes interspecific competition doesn't just affect the population **size** of a species, it totally **prevents** the species living in an area at all.

Since the introduction of the **grey squirrel** into Britain, the native **red squirrel** has **disappeared** from large areas because of interspecific competition. In the few areas where both species still live, both populations are **smaller** than they would be if there was only one kind of squirrel there.

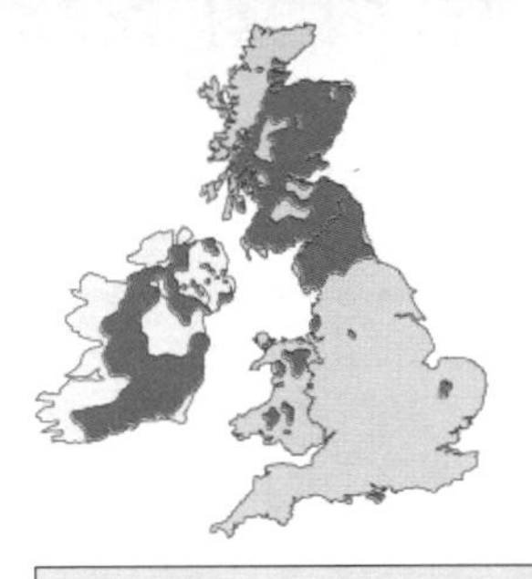

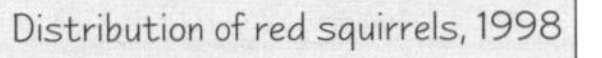

Distribution of red squirrels, 1998

Distribution of grey squirrels, 1998

maps courtesy of Forest Research

Predator and Prey populations are interlinked

The presence of **predators** affects the **size** of populations. The **graph** shows **population fluctuations** of the **lynx** and its prey, the **snowshoe hare**, in Canada — it's a good example of how closely related predator and prey numbers are.

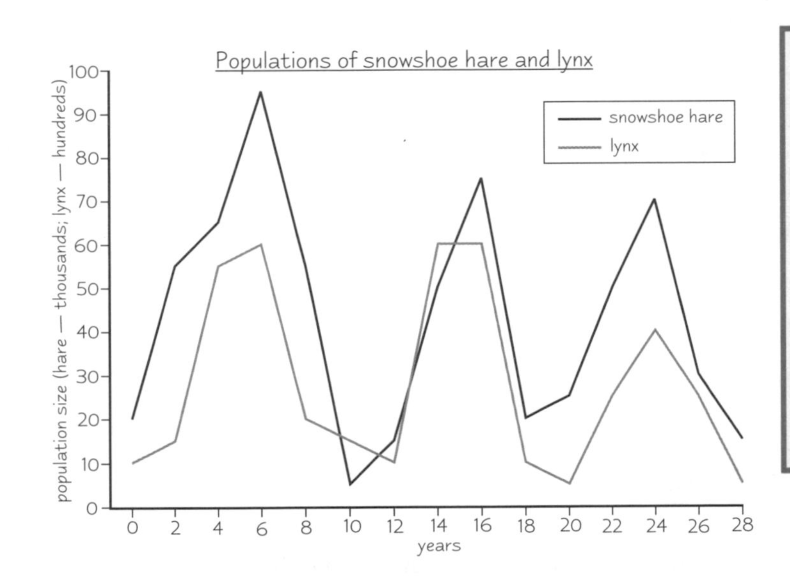

1) The theory is that as the **population** of **prey grows**, there is **more food for predators**, and so the predator **population grows**.
2) As the **predator** population gets **larger**, **more prey** is **eaten** so the **prey** population **falls**.
3) Then, there is **less food** for the **predators** so their **numbers go down**, and so on.
4) In nature you rarely get this **exact pattern** because it's unusual to get a situation where a species is eaten by just **one** predator, or a predator just eats **one** prey species.

Practice Questions

Q1 Define the term 'community'.

Q2 Give one example of a density-dependent factor.

Q3 Explain how the populations of a predator and its prey change over time.

Q4 Name five abiotic factors that can affect population size.

Exam Questions

Q1 Explain the difference between intraspecific and interspecific competition. [2 marks]

Q2 Factors that regulate population size can be density dependent or density independent.
Explain the difference between these two types. [4 marks]

Lack of chocolate is a revision limiting factor...

Again there's some stuff you've seen lots before here — like predator-prey relationships, but it's jumbled up with new stuff too. So don't be fooled into thinking, "yeah yeah, been there, done that, got the T-shirt — in fact I invented the T-shirt me, this A2 biology lark's a cinch, reckon I might just sack this revision off"... you get the point, this stuff needs learning.

Controlling Pests

We had a pest in our cupboard once. He was a mouse and he ate all my rice so we called him Paddy.

There are Three methods of Pest Control

Pests are species that negatively affect human activities. They can be **weeds** that out-compete crops, **insects** that carry diseases or **fungi** that reduce crop yields. They could also be animals such as **rabbits** or **deer** that eat crops. There are three main approaches to dealing with pests:

1) **Chemical control** uses chemicals like pesticides to kill the pest species.
2) **Biological control** uses living organisms to control the pest. These include **predators**, **parasites** and **pathogens** which kill the pest or prevent it from reproducing.
3) **Integrated pest management** mixes both chemical and biological control together to combat pests.

Chemical Pesticides are pretty Effective

1) **Herbicides** are used to combat weeds. There are two kinds:
   - **Contact herbicides** kill weeds when they are sprayed onto their **surface**.
   - **Systemic herbicides** have to be **absorbed** into the weed to kill it. High doses of synthetic auxins (plant hormones) are used as systemic herbicides.
2) **Fungicides** are used to fight fungal infections. Like herbicides they can be either contact or systemic.
3) **Insecticides** are used against... guess what... insect pests. There are three types this time:
   - **Contact insecticides** have to come into direct contact with the insect.
   - **Systemic insecticides** are absorbed by the plant and carried in its **phloem**. They kill the insects that feed on the plant's sap.
   - **Stomach ingestion insecticides** are sprayed over the crop and are **consumed** when pests eat the plants.

Chemical Pesticides can cause Problems

Some pesticides are **persistent** — that means that they don't break down in the environment or within the tissues of living organisms. This causes big problems for species that aren't supposed to be damaged by the pesticide:

Bioaccumulation happens when persistent pesticides stay in an organism's living tissue. As the organism consumes more and more of the pesticide the levels in its tissues become **increasingly toxic**. The pesticides are passed from one **trophic level** to the next, and each time they become increasingly **concentrated**. The species that are highest up in the food chain (e.g. birds of prey) can end up with **lethal** concentrations in their bodies.

Pesticides that aren't specific cause lots of problems because they kill other species, as well as the ones that they're supposed to attack.

Biological Control can be Better for the Environment

Biological control avoids the need for putting extra **chemicals** into the environment. Organisms that are used in biological control include:

1) **Insect parasites** which are specific to the pest. Most lay their **eggs** on or in the host and then when the eggs hatch the larvae eat the host from the **inside**. Nice...
2) **Predators** are **carnivorous** species used to control insects and other animal pests.
3) **Pathogens** are **bacteria** and **viruses** used to kill pests. For example, the bacterium ***Bacillus thuringiensis*** produces a **toxin** that kills a wide range of **caterpillars**.

Controlling Pests

Biological control has **Advantages** and **Disadvantages**

Advantages

1) The control organism is usually **specific** and only the pest species will be affected.
2) **No chemicals** are used, so problems of bioaccumulation and pollution are avoided.
3) Control organisms usually establish a population so there is **no need** for **re-application**.
4) Pests don't usually develop genetic resistance to the biological control agent (this is a big problem with chemical fertilisers).

Disadvantages

1) Control is **slower** than using chemical pesticides because you have to wait for the biological control agent to establish a large enough **population** to control the pests.
2) Generally biological control **doesn't permanently exterminate** the pest — instead it's reduced it to a level where it is no longer a big problem.
3) It can be **unpredictable** — it's really hard to work out what all the **knock-on effects** of the introduction of the biological control species will be.
4) A lot of **scientific research** into the relationship between the pest species and the biological control agent is needed. This research takes **time** and costs **money**.
5) If the control species has several sources of food its population levels might **grow** and it may **become a pest** itself.

There are **Other Ways** of dealing with **Pests**

In recent years farmers have started to use **integrated pest management** where chemical and biological methods are combined.

Biological control is used to keep the pest down so that it doesn't affect the **profitability** of the farm. If a pest **outbreak** occurs then the farmer uses a specific pesticide for a **short** period of time.

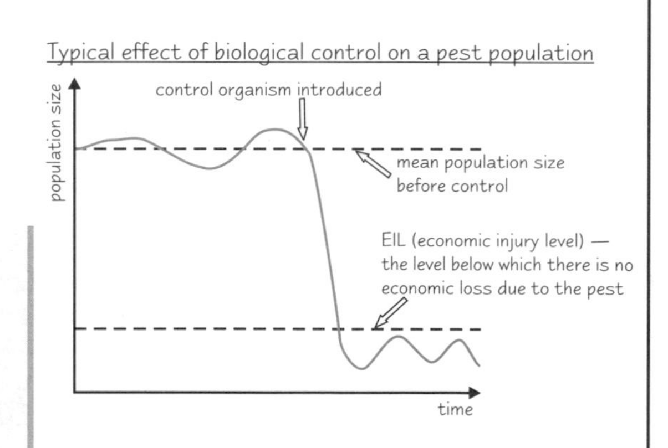

Crop rotation can help control pests — changing the crop each year makes it difficult for pests that can only feed on one species to become established.

Genetic modification can be used to produce crops which are **resistant** to pests, so no pest control is required. (Although it is possible for pests to evolve so they can feed on the genetically modified organisms.)

Practice Questions

Q1 What is the difference between a contact pesticide and a systemic pesticide?

Q2 What problem is caused in food chains by persistent pesticides?

Q3 What is integrated pest management?

Exam Questions

Q1 A chemical company is developing a new weed killer. State three features that the company should aim to include in their herbicide to ensure that it causes as little environmental damage as possible. [3 marks]

Q2 Explain why some farmers prefer to use pesticides rather than biological control to deal with pests. [5 marks]

There's no need to be pestimistic, this stuff's easy...

When you realise how difficult it is to control pests you can understand why some people think that genetic modification is the way forward. The thing that worries people is that, like biological control, it's hard for scientists to know exactly what the consequences of growing GM foods will be in the long term. It could be a great thing, or it could be a scary mistake.

Managing Ecosystems

These pages look at how ecosystems are affected by humans and how they can be managed for the good of both humans and the natural environment.

Intensive Farming methods can Affect the environment

Intensive farming methods (producing large amounts of food from relatively small areas of land) have **boosted food production**, stopping people starving in some parts of the world and making sure there is always plenty of good quality food on supermarket shelves. **But** these methods can cause **environmental problems**:

- **Hedgerows** are **removed** to allow access for large machines or to enlarge fields. This **destroys** the **habitat** of many species and reduces species diversity.
- Large amounts of **inorganic fertiliser** are used. This can **leach** through the soil and **pollute** nearby streams and rivers, killing many species.
- '**Factory farming**' methods of keeping many animals in a small space produce large amounts of **waste**, which can cause pollution if not disposed of carefully.
- Large scale use of **pesticides** can result in water **pollution** and may also kill non-pest species. Food can become contaminated.

Farm Land can be Managed to Increase Biodiversity

Here are a few examples of things that farmers can do to increase biodiversity on their land:

1) **Planting hedgerows** — This increases **species diversity** and some of the species attracted might well feed on potential pests, helping the farmer. Habitat variety can be further increased if the farmer leaves small areas of woodland on their land.
2) **Intercropping** — This is the practice of growing two or more crops in the same field at the **same time**. It can produce a greater yield on a given piece of land, by using space that would otherwise be wasted with a single crop. Careful **planning** is required, taking into account the soil, climate, crops and varieties. An example is planting a deep-rooted crop with a shallow-rooted crop to make maximum use of soil nutrients. Intercropping encourages **biodiversity** because more species can be supported by two different crops than just one. It can also reduce pests — each crop may contain a chemical that repels the pest species of the other.
3) **Maintain** existing **ponds** and **wetland** rather than draining them.
4) **Create natural meadows** (a natural meadow has a high diversity of plant species) and produce hay rather than grass meadows (that contain just grass) that create silage (an animal feed that's like hay).

Woodland can be Sustained by Good Management

Today, most woodland in the UK is used for either timber production or leisure.
Woodland is managed so that it can continue to provide habitats for wildlife.

Timber Production

Wood that is grown for timber is **deliberately** planted — species such as **pine** are used because they **grow quickly**. As soon as one crop of trees has been felled, more are planted.

Coppicing is a **traditional** method of timber production. Trees like hazel and sweet chestnut are cut down to their base so that they begin to sprout — each shoot is harvested when it matures. Coppicing produces lots of long poles without the trees themselves having to be felled.

Leisure

Management for leisure use involves the creation of **mixed woodland** with paths and open spaces. Planting native shrubs like blackthorn, hawthorn and holly in shady parts of the wood can provide excellent habitats for birds such as wrens.

Dead wood is left where it is as it provides food and shelter for **fungi** and **invertebrates** like woodlice and wood-boring beetles. These are the foundation of many **food chains**, attracting many birds e.g. woodpeckers, nuthatches, bats and small mammals.

Managing Ecosystems

Grassland needs Management too

In Britain, grassland is mainly used for livestock or parkland. You might think that a field of grass grazed by cows or sheep just sorts itself out, but it actually needs fairly careful management.

Grazing needs to be carefully controlled. Sheep and dairy cattle prefer short grass, but beef cattle do better when grass is slightly longer. The farmer needs to rotate the stock at intervals to ensure that pastures are evenly grazed. Grazing also stops shrubs from growing which can shade the grass and stop it growing so well.

Mowing is used to maintain areas of parkland. Like grazing, it stops the development of larger plant species that would eventually displace the grass.

If mowing or grazing is inadequate, **manual shrub clearance** of the grassland is sometimes needed.

Amazingly, even **fire** can be used for the management of grassland. Perennial grasses have growing points which are below the soil, so they can survive fire. Fire can improve grass quality, quantity and palatability (how tasty it is) — it removes dead plant material, increases new growth and controls competing weeds.

The European Union has set up the Natura 2000 Project

In 1992, the EU Habitats Directive set up **Natura 2000** to ensure that sensitive **habitats** and **endangered species** across Europe are **conserved**. Each member state was asked to provide a list of sensitive sites and areas where threatened species were found, and to agree a plan of action to manage them.

Practice Questions

Q1 How does intensive farming affect the environment?

Q2 Explain how farmers could increase biodiversity on their land.

Q3 How does mowing help manage grassland?

Q4 Why do perennial grasses survive when fire sweeps the land?

Exam Questions

Q1 Describe ways in which woodland can be managed for sustained timber production. [4 marks]

Q2 The graph shows the pattern of grass growth and the requirements of livestock living off it, throughout a year.

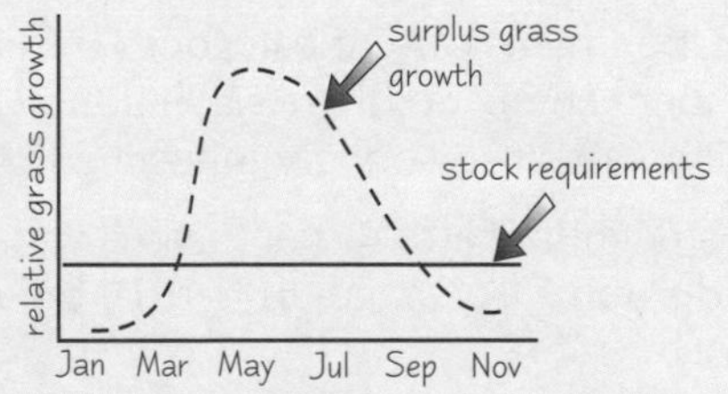

(a) Suggest what measures would be needed to maintain the grassland during May to September. [1 mark]

(b) Suggest how the stock might be properly fed between September and the following May. [1 mark]

(c) Land managers are sometimes asked to leave a strip around 2m wide round the edge of their fields, where the grass is left to grow long. Suggest a possible environmental benefit of this. [1 mark]

Natura 2000 — isn't that a nudist colony...

There's something so appealing about the idea of running free in the great outdoors, feeling the wind in your hair and the sun on your back. How I long to be out there with the little bunny rabbits and the birds and the little fishies, communing with nature. With my clothes on, of course. Now enough of this nonsense — there's biology to learn here, y'know.

Variation

Ever wondered why no two people are exactly alike? No, well nor have I, actually, but it's time to start thinking about it. This variation is partly genetic and partly due to differences in the environment.

Variation can be Continuous or Discontinuous

Discontinuous variation

This is when there are two or more **distinct types**, and each individual is one of these types, for example:

Sex — you're either male or female

Blood group — you can be group A, group B, group AB or group O, but no intermediates.

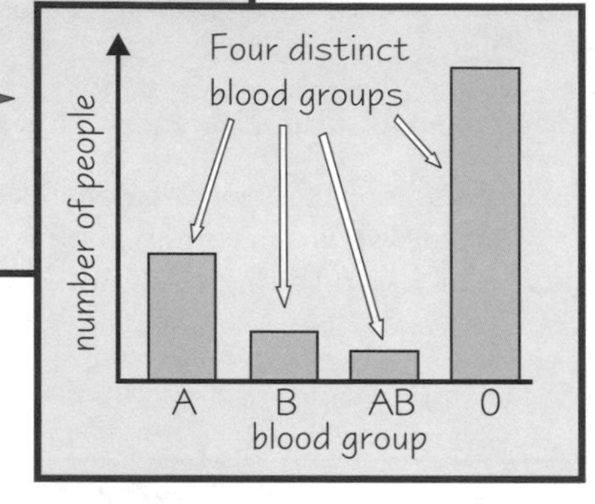

Continuous variation

This is when the individuals in a population vary along a **range**, with **no distinct types**, for example:

Height — you could be any height over a range

Weight — you could be any weight over a range

Skin colour — any shade from very dark to very pale

Hair colour — any shade from blonde to black

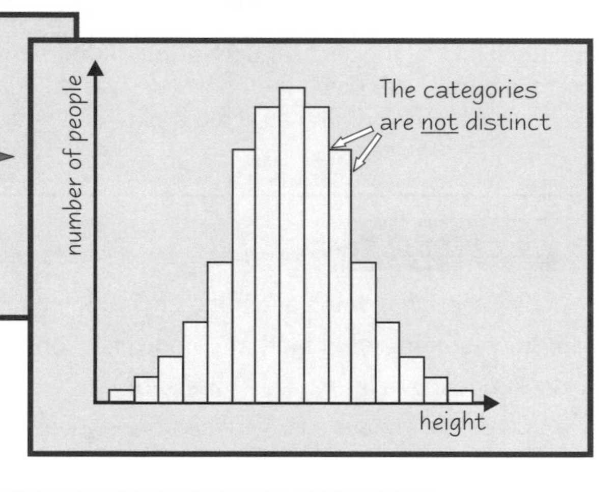

Discontinuous Variation is influenced by One or a few Genes

Discontinuous variation has clear-cut categories because it depends on only one or a few genes (it is monogenic). So, there's a **limited number** of possible phenotypes (see next page). Discontinuous variation isn't so strongly influenced by the environment.

Because it's pretty straightforward, discontinuous variation is the kind of variation usually studied in genetic diagrams (see pages 100-101), but most variation in real life is actually continuous variation.

Continuous variation is Polygenic

Polygenic means that several genes affect the same characteristic.
Continuous variation can be more strongly affected by the **environment**. Because of the interaction of loads of different genes plus the effect of the environment, there's lots of possible **phenotypes**.

Example:
Human body mass shows **continuous variation**. Your mass is **partly genetic** (big parents often have big children), but body mass is also strongly affected by **environmental factors** like diet and exercise.

Variation

You need to Learn some Key Terms

There's a lot of **fancy words** used about genetics and inheritance.
Before we go any further make sure you know all of these:

chromosome — A strand of **genetic material** (DNA) found in the nucleus of a cell. Each chromosome consists of one **molecule** of **DNA** and **histone proteins**.

chromatid — **One** of the two **identical strands** of genetic material that make up a chromosome during cell division.

homologous — Homologous chromosomes are **a pair of equivalent chromosomes** with the **same structure** and **arrangement** of genes — usually one is inherited from the mother and one from the father.

bivalent — A **pair** of **homologous** chromosomes.

haploid — A cell with only **half** the number of chromosomes of the parent organism (only **one copy** of each chromosome), e.g. **sperm** and **egg** cells.

diploid — A cell with the **full number** of chromosomes — in **pairs of homologous chromosomes**.

gene — A **section** of **DNA** on a chromosome which controls a **characteristic** in an organism. It carries the **genetic code** to make one or more **polypeptide** or **protein**, or to make **RNA**.

locus — The **position** on a chromosome where **a particular gene** is located.

allele — An **alternative form** of a gene. E.g. in pea plants, the gene for height has **two forms** — one allele for tall plants and one allele for short plants.

genotype — The **alleles** a particular individual has.

phenotype — An individual's **characteristics**, e.g. eye colour, blood group.

homozygous — An individual with **two copies** of the **same allele** for a particular gene.

heterozygous — An individual with **two different alleles** for a particular gene.

dominant — The condition in which the effect of only **one allele** is **apparent** in the phenotype, even in the presence of an alternative allele.

codominance — The phenomenon in a **heterozygote** in which the effects of **both alleles** are **apparent** in the phenotype.

recessive — The condition in which the effect of an allele is **apparent** in the phenotype of a **diploid** organism only in the presence of another **identical allele**.

linked — **Genes** located on the **same chromosome** that are often **inherited together**.

Practice Questions

Q1 Define 'discontinuous variation'.

Q2 Give two examples of characteristics that show continuous variation, and two that show discontinuous variation.

Q3 Define the terms 'genotype' and 'phenotype'.

Exam Questions

Q1 Compare characteristics showing continuous and discontinuous variation with reference to:

a) the extent to which they are affected by the environment. [1 mark]

b) the number of genes that control them. [1 mark]

Q2 Explain the difference between a homozygous and a heterozygous individual. [2 marks]

Variety is the spice of... genes and the environment...

By now you should have a pretty good idea which traits show continuous variation and which ones show discontinuous variation. It's amazing to think of how many genes influence the way that we look and behave. It's the reason we're all so lovely and unique... my parents said they were glad they'd never have another child quite like me — I can't imagine why.

Genes, Environment and Phenotype

The fact that you're such a fantastic person is partly because of the genes you got from your parents, but it's also because of the way they brought you up. Either way, you can thank them for it.

Genes affect Phenotype Variation

There are two main sources of genetic variation — sexual reproduction and meiosis. This results in phenotype variation.

Sexual Reproduction

Sexual reproduction is where the gametes from two individuals **randomly fuse** at fertilisation. This process mixes up the alleles in new combinations, creating more **genetic variety** in a species. This means that **survival** (of at least some individuals in a population) is more likely, so there's less chance of the species becoming extinct. If all individuals were the same, one set of environmental conditions or one disease could easily wipe them **all** out.

Meiosis

Meiosis is a special kind of **cell division**. It **halves** the chromosome number. It's used for sexual reproduction in plants and animals. In animals, meiosis produces the **gametes** (sperm and egg cells) and takes place in the **testes** and **ovaries**. When the **gametes** fuse at **fertilisation** they combine their chromosomes, so the chromosome number is **restored**. This produces new combinations of alleles, creating **genetic variety**.

During meiosis variety can be produced in three different ways:

If you can't remember what goes on in meiosis have a look back at your AS notes.

1) **Independent assortment** of chromosomes

During **meiosis I** (the first of two divisions), the pairs of homologous chromosomes **separate**. The chromosomes from each pair end up **randomly** in one of the new cells, so you can get **different chromosome combinations**. In **meiosis II**, there's also random assortment of **chromatids**.

One pair of chromosomes would give **2** different types of haploid cell.

- Two pairs would give 2^2 possible haploid cells = **4 possibilities**.
- 23 pairs, like in humans, give 2^{23} possible haploid cells = over **8 million possibilities**. (Your parents would have to have millions of children before they stood any chance of having two genetically the same — unless they have twins.)

2) **Crossing Over**

Chromosomes often swap parts of their chromatids during **meiosis I**. This creates **new combinations** of alleles on those chromosomes (called recombinant chromosomes), separating alleles that are normally inherited together (**linked**).

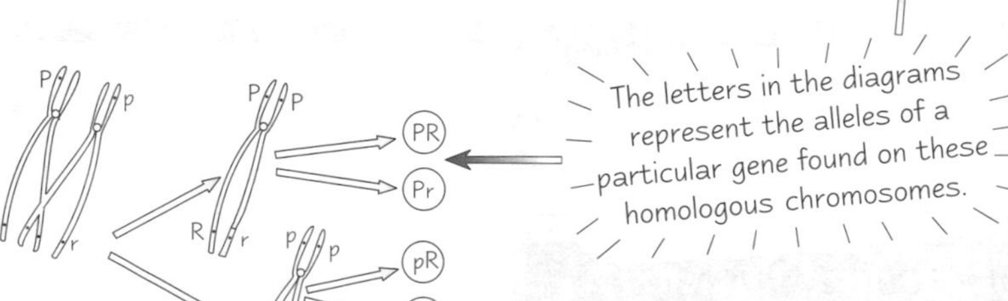

3) **Mutation**

Mistakes sometimes happen during cell division, producing a **completely new** characteristic (see pages 98-99).

Environment affects Phenotype too

A lot of variation in characteristics (phenotype) is due to differences in genotype, but **environment** also has an effect:

1) People are on average much **taller** today than they were 200 years ago (if you're a strapping six-footer, you'll probably bump your head on the ceilings of an old house). This is thought to be because our diet is much better.
2) **Plant growth** is strongly affected by the environment — plants show better, healthier growth when there are more **nitrates** and other minerals available in the soil.

Genes, Environment and Phenotype

Genes and Environment interact in the Phenotype

Pea plants provide a clear example of the **interaction** between genes and environment that produces a **phenotype**.

Pea plants come in tall and dwarf forms.
This characteristic (tall or dwarf) is passed on from one generation to the next, so we can tell that it is **genetic**.

However, the tall plants vary in height, and so do the dwarf plants, so **environment** is involved too.

Tall or dwarf is discontinuous variation. Height variation among the plants of each type is continuous variation.

The F_1 generation is the first set of offspring form two parents. If you then breed these offspring together, you produce the F_2 generation.

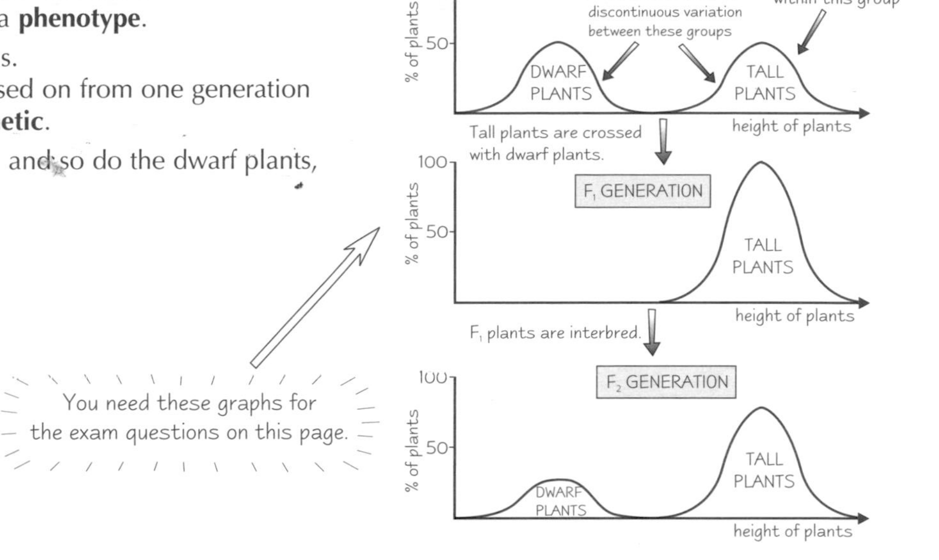

Twins can show the effects of Genes and Environment

Studies of **twins** have been used to find out if a human characteristic is mainly influenced by **genes** or by **environment**.

Monozygotic twins ("identical twins") have **identical genes** and **alleles**, because they both developed from the **same fertilised egg**. This means that if there are any differences in their characteristics, they may be due to the **environment**. Occasionally, monozygotic twins are raised **separately**, and comparing differences between them (compared with twins raised together) could show how important these environmental influences are.

Practice Questions

Q1 State three ways in which meiosis helps to create variety within a species.

Q2 Describe the relationship between the effect of genetic and environmental variation on phenotype.

Q3 Why don't monozygotic twins always have identical characteristics?

Exam Questions

Q1 If the body cells of an organism contain three pairs of chromosomes, how many different chromosome combinations can be produced in the gametes of this organism as a result of independent assortment? [2 marks]

To answer these questions, look at the data for the heights of pea plants in the graphs on this page.

Q2 How does the information about the parent plants suggest that:
a) height is partly genetically determined? b) height partly depends on the environment? [2 marks]

Q3 Do the heights of the F_1 generation show continuous variation, or discontinuous variation, or both? Explain your answer. [2 marks]

Q4 Compare the heights of the parental generation with the F_2 generation.
How does this comparison support the idea that height is partly genetically determined? [1 mark]

Having an identical twin is like having a clone...

That's quite a weird thought... if you've got same-sex twins in your class, it's interesting to investigate how similar they are. How can you tell them apart? If they're identical twins, it must be an environmental difference, e.g. one's dyed their hair pink, or has a tattoo. It's often hard to tell if they're really identical though, or just fraternal twins who look very similar.

Mutation and Phenotype

Mutation sounds quite exciting, but if you're expecting pictures of chickens with two heads or green monsters, you're going to be disappointed. Anyway, here's what happens when cell division goes wrong.

Point Mutations are Changes in the DNA Base Sequence

Before a cell divides, its DNA is replicated (copied) — look at your AS notes for more detail on DNA replication. Sometimes the base sequence of the DNA (the genetic code) gets changed. This is a **gene mutation** and it can make the DNA code for a different protein. A change of **one base** (C,G,A or T) is called a **point mutation**. The effect of a point mutation depends on exactly what happens:

No mutations

C G A — G T T — G C A

This forms triplet codes for these amino acids:

DNA code	amino acids
CGA	alanine
GTT	glutamine
GCA	arginine

substitution here

C T A — G T T — G C A

A substitution changes one triplet code and usually one amino acid like this:

DNA code	amino acids
CTA	aspartic acid
GTT	glutamine
GCA	arginine

Substitution

One base is **swapped** for another in a triplet code. It means that the gene will make a **similar protein** to the normal protein, but with just **one amino acid** different. Because the structure of a protein is so important, this can have a big effect. The **sickle-cell** allele is the result of a base substitution. (See page 101 for more on sickle-cell anaemia)

Insertion (addition)

An **extra** nucleotide (with a base) is included in the DNA molecule. This has a much bigger effect than a substitution, because it causes **all** the following triplet codes in the gene to be altered. If there's an insertion, the gene doesn't make **any** useful protein at all, which can cause serious problems.

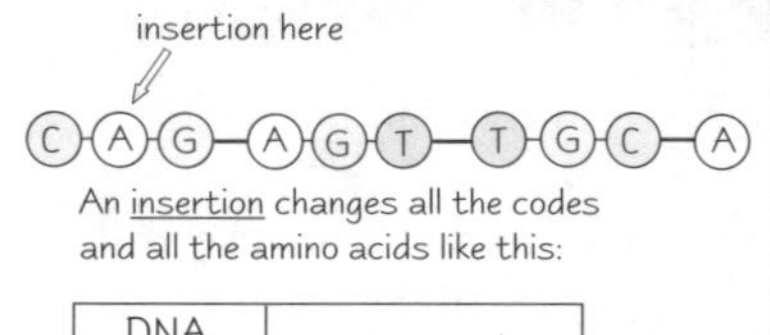

An insertion changes all the codes and all the amino acids like this:

DNA code	amino acids
CAG	valine
AGT	serine
TGC	threonine

Deletion

A nucleotide (with its base) is **missed** out when the DNA is copied. Like insertion, this **shifts along** all the triplet codes after it, so it really messes things up. This is also known as a '**frame shift**' mutation.

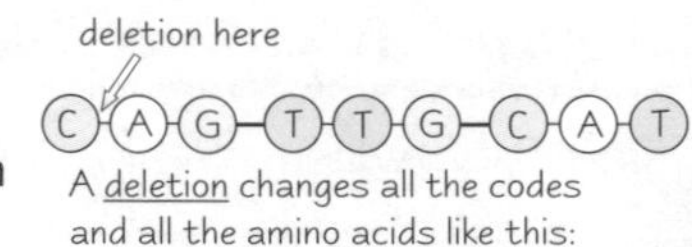

A deletion changes all the codes and all the amino acids like this:

DNA code	amino acids
CAG	valine
TTG	asparagine
CAT	valine

Chromosome Mutations are caused by Errors in Cell Division

A **chromosome mutation** is where one or more **whole chromosomes** (or **parts** of chromosomes) get in the **wrong place** during cell division. If this happens during meiosis, then the chromosome abnormality will be in the **gametes**, so it gets passed to the **next generation**. There are **three main types** of chromosome mutation:

Polyploidy

All the chromosomes fail to segregate during meiosis. This is called **non-disjunction**. This creates a diploid gamete (with **two sets** of chromosomes) and, after fertilisation, the zygote will have **three** (or even more) complete sets of chromosomes.

Polysomy

Non-disjunction of some, but not all, of the chromosomes. E.g. a gamete may have just one extra chromosome, so the offspring also has one extra chromosome in each cell. About **95%** of **Down syndrome** cases in humans are caused by a person having an extra copy (**trisomy**) of chromosome 21.

Translocation

This is where **part** of a chromosome gets broken off and then **reattaches** in a different place. This doesn't usually have such a big effect on the phenotype as polyploidy or polysomy, but it can still have serious consequences. About **5%** of **Down syndrome** cases are caused by this.

Mutation and Phenotype

Mutagens make Mutations more likely

Mutations are accidents, and can happen for no obvious reason.
However **mutagens** make mutations more likely to happen:

1) **Radiation**
Some types of **radiation** are mutagenic. This includes X-rays, UV rays and ionising radiation such as gamma-rays.

2) **Chemicals**
Some **chemicals** are mutagens. Most of these chemicals are also carcinogens (they cause cancer).
E.g. mustard gas and substances in tobacco smoke. These chemicals cause **point mutations** in the DNA.
Other chemical mutagens affect the structure of **chromosomes**, making chromosome mutation more likely.
E.g. the dye **colchicine** has this effect and is used by **plant breeders** to produce new plant species.

Mutations are often Harmful

Any random change to the DNA in a cell is probably going to be **damaging**.
Most **genetic diseases** are the result of mutations.

Changes in the cell's DNA might mean that it codes for a different protein, or for none at all, which could stop an important process from working.

The Human Genome Project (HGP)
This international project to **map the positions** of different genes on the chromosomes was completed in 2003.
Information from the HGP will help scientists understand **mutations** and **genetic diseases**.

Albino wallaby Mervyn gets upset when people call him a mutant.

Mutations can lead to Evolution

Occasionally a mutation creates an **improvement**. If so, the mutant will have a selective advantage and will probably end up having more offspring, so **natural selection** causes the mutant form to become more common.
This type of mutation is important in **evolution**.

Practice Questions

Q1 Name the three main types of gene mutation.
Q2 Explain the difference between polyploidy and polysomy.
Q3 Give examples of two chemical mutagens.

Exam Questions

Q1 Explain why the deletion of three adjacent nucleotides in a gene mutation will usually have a less severe effect on the phenotype than the deletion of one nucleotide. [3 marks]

Q2 Suggest explanations for these facts:
a) Radiographers in hospitals stay behind a lead screen when giving X-rays to patients. [2 marks]
b) Excessive exposure to bright sunlight can cause skin cancer. [2 marks]
c) When fruit-flies are exposed to X-rays and then mated, some of their offspring have abnormal white eyes or deformed wings. [2 marks]

So you're telling me I'm a mutant...

Loads of genetic diseases start off as just a random mutation in one person, then the mistake just keeps getting passed on down the generations. On the other hand, a mutation could be the reason for your stunning good looks.

Inheritance

Brace yourself for two pages of genetic diagrams. You need to get comfortable with these because in the exam you'll not only have to interpret them, you might have to draw some of your own. It's probably a smart idea to learn some of the most common patterns and ratios, then you'll be able to apply them to new examples in the exam.

Monohybrid Inheritance Involves One Characteristic

Each individual has **two copies** of a gene. But they **segregate** when the sex cells are formed in meiosis, so each **gamete** contains only **one copy** of **every** gene. Monohybrid inheritance is the **simplest** form of inheritance — it's just inheritance where a single gene is being considered. A **monohybrid cross** is a genetic cross for only one gene:

Example

In fruit flies, the allele for **normal wings** is **dominant** (N), and the allele for **vestigial** (short) wings is **recessive** (n).

A normal-winged fruit fly is crossed with a fruit fly that has vestigial wings. **All** the offspring are normal-winged. These flies then **interbreed**, and the next generation shows a **3:1 ratio** of normal wings to vestigial wings, i.e. a 75% chance of normal wings and a 25% chance of vestigial wings.

The first set of offspring from an experiment like this, where the two parents are true-breeding (homozygous), is called the F_1 generation. If you then breed these offspring together, you produce the F_2 generation.

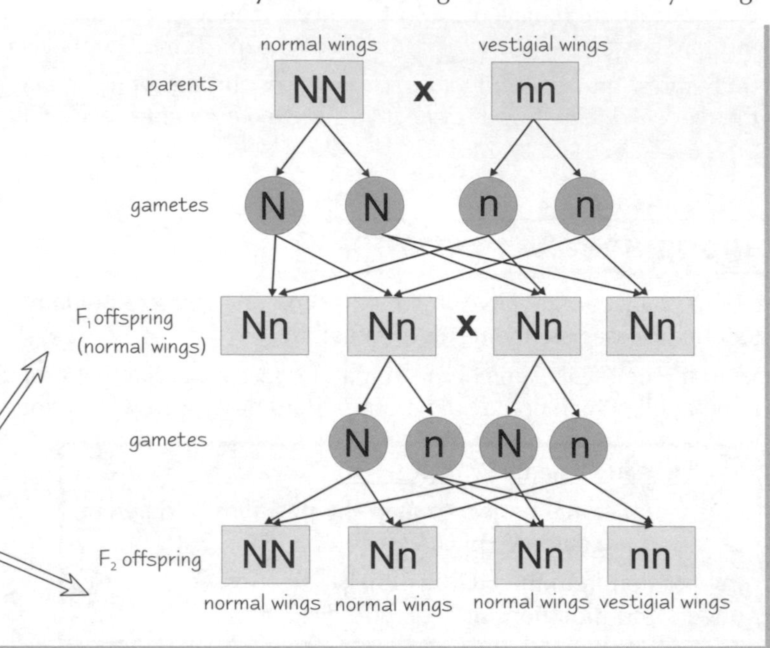

Genetic diagrams like this show all the possible combinations of gametes from the parents. Remember to use a capital letter for a dominant allele, and a small letter for a recessive allele.

A Test Cross helps you find out an Individual's Genotype

Sometimes you might cross a normal-winged fly with a vestigial-winged fly and get a **1:1 ratio** of normal wings to vestigial wings in the offspring, instead of all normal. This happens if the normal-winged fly is **heterozygous** — it has **one allele** for **normal** wings, and **one allele** for **vestigial** wings. Because the allele for vestigial wings is recessive, it doesn't show up in the phenotype of heterozygous flies — vestigial wings is a recessive condition.

Compare this with the first diagram on this page. In each case, the normal fly **looks** the same (they have the same **phenotype**). The only way of telling its genotype is by a **breeding experiment** where you mate it with a recessive individual — in this case that's a fly with **two alleles** for **vestigial** wings (remember this because it'll crop up in the exam). This is called a **test cross**.

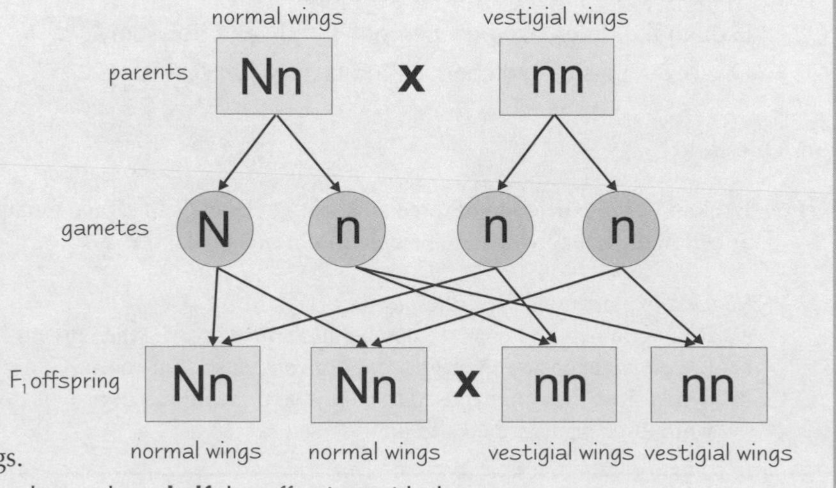

- a **homozygous** normal winged fly produces **all** offspring with the **dominant** characteristic when it's crossed with a fly with vestigial wings.
- a **heterozygous** normal winged fly produces about **half** the offspring with the **recessive** characteristic when it's crossed with a fly with vestigial wings.

Inheritance

Alleles can be Codominant

Occasionally, alleles show **codominance**.
One example in humans is the allele for **sickle-cell anaemia**:

- Normal people have two alleles for normal haemoglobin (**H^NH^N**).
- People with **sickle-cell anaemia** have two alleles for the disease (**H^SH^S**). They have abnormal haemoglobin, which makes their red blood cells sickle-shaped and unable to carry oxygen properly. Sufferers usually die quite young.
- Heterozygous people (H^NH^S) have an in-between phenotype, called the **sickle-cell trait**. Some of their haemoglobin is normal and some is abnormal, but the red blood cells are normal-shaped. The two alleles are **codominant**, because they're **both** expressed in the **phenotype**.
- The sickle-cell allele is a result of a **mutation** (see page 98).

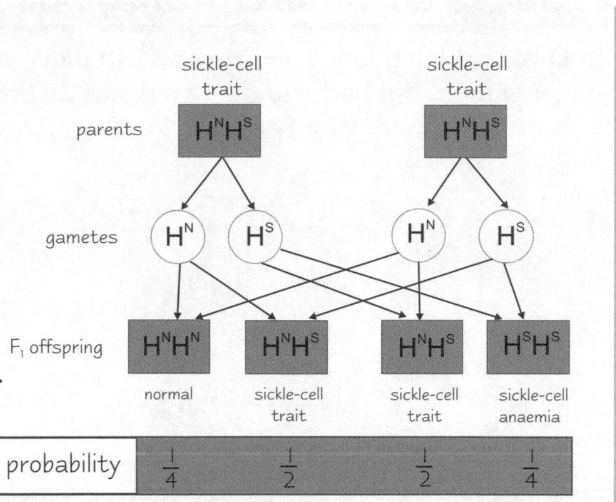

Some Genes have Multiple Alleles

Inheritance is more complicated when there are more than two alleles of the same gene — **multiple alleles**.
E.g. in the **ABO blood group system** there are **three alleles** for blood type:

I^O is the **recessive** allele for blood group **O**. **I^A** is the allele for blood group **A**. **I^B** is the allele for blood group **B**.

Alleles I^A and I^B are **codominant** — people with copies of **both** these alleles will have a **phenotype** that expresses **both** alleles, i.e. blood group **AB**. In the diagram below, if members of a couple who are both **heterozygous** for blood groups A and B have children, those children could have one of **four** different blood groups — A, B, O or AB.

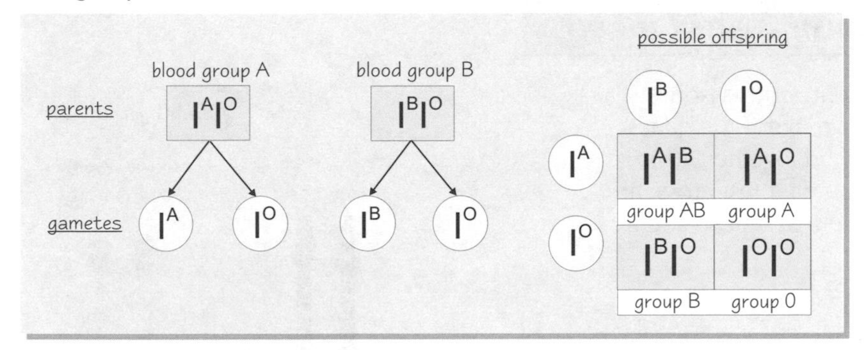

Recessive blood group alleles are normally really rare, but it just so happens that loads of people in Britain descended from people who were $I^O I^O$, so it's really common.

Practice Questions

Q1 What genetic ratios do you expect from each of these crosses?
a) Aa × Aa b) AA × Aa c) Aa × aa

Q2 What is sickle-cell anaemia? What type of inheritance pattern do sickle-cell alleles show?

Q3 Do the genetics of the ABO blood group system show multiple alleles, codominance, or both?

Exam Questions

Q1 List the six possible genotypes for the human ABO blood groups. [3 marks]

Q2 In pea plants, the condition purple flowers is dominant over the condition white flowers.
How would you find out if a purple-flowered plant is homozygous or heterozygous? [3 marks]

It's hard to do test crosses on humans...

If you're wondering whether you're heterozygous for a particular trait, it's probably not an option to breed with a recessive person, and then have lots of babies and see what they look like, unless you take your science homework very seriously.

Inheritance

There's so much to say about inheritance that we've generously stuck in another two pages for you to enjoy.

Genes on *Different Chromosomes* Segregate *Independently*

Dihybrid inheritance shows how **two** different unlinked genes are inherited. Each gene gives a 3:1 ratio in the F_2 generation, but because the two genes do this **independently**, it makes a **9:3:3:1 ratio** overall. This diagram shows how this happens for two traits in the fruit fly.

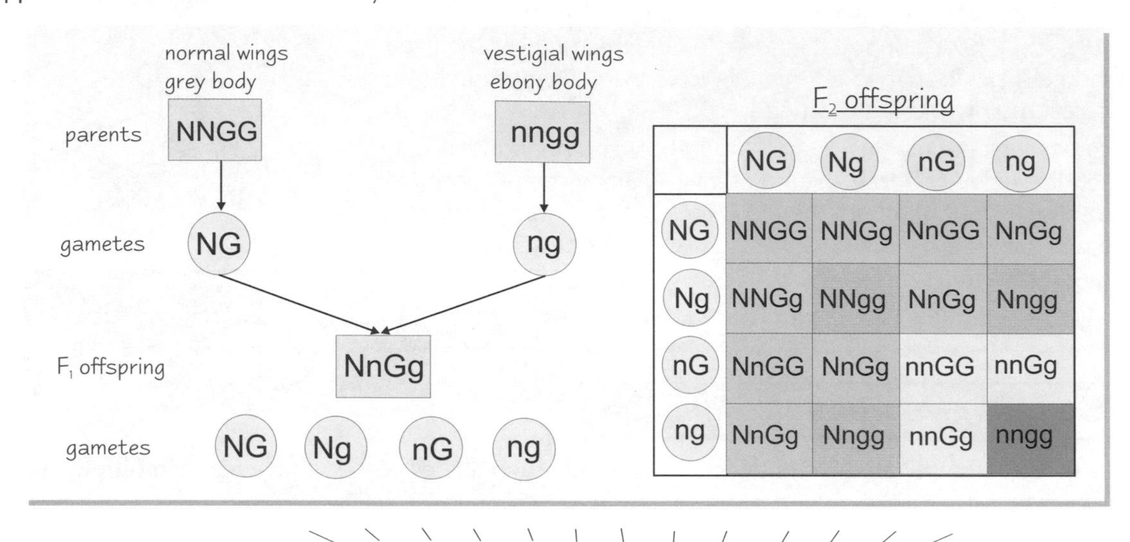

	NG	Ng	nG	ng
NG	NNGG	NNGg	NnGG	NnGg
Ng	NNGg	NNgg	NnGg	Nngg
nG	NnGG	NnGg	nnGG	nnGg
ng	NnGg	Nngg	nnGg	nngg

Crossing an F_1 fly with a double recessive fly (vestigial wings and ebony body) gives a 1:1:1:1 ratio. Check your understanding by working this out yourself.

Genes on the *Same Chromosome* are *Linked*

The crosses so far all assume that two different genes assort (**segregate**) independently. This is usually true, **but** if the genes are on the same chromosome, they're inherited together. This is called **linkage**. If genes are **linked**, when you do a test cross, an **expected** 9:3:3:1 ratio will turn out more like a 3:1 ratio, and a 1:1:1:1 ratio turns out more like 1:1.

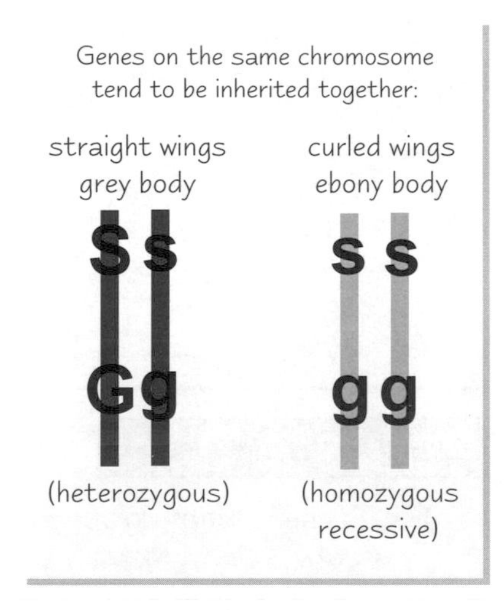

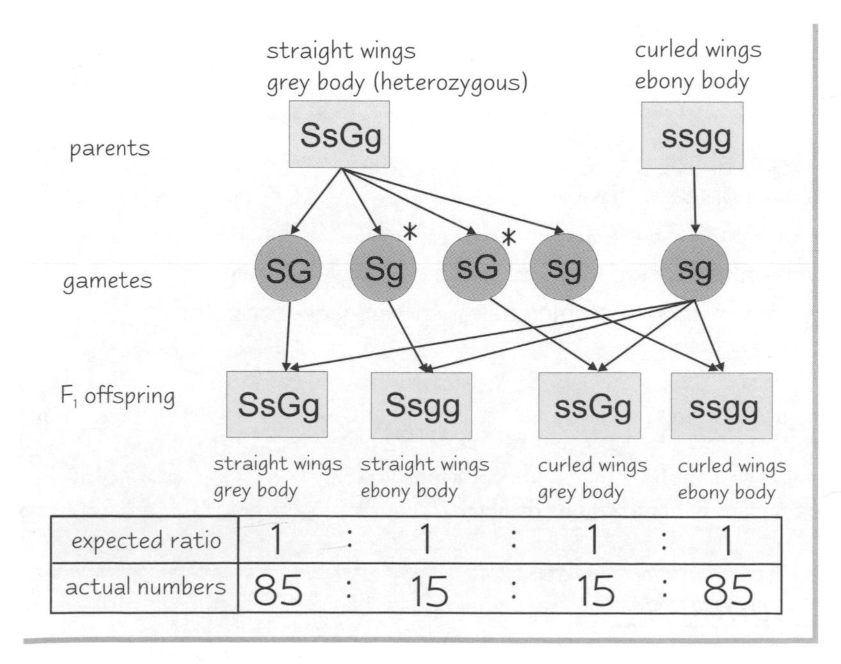

	SsGg	Ssgg	ssGg	ssgg
expected ratio	1	1	1	1
actual numbers	85	15	15	85

It's not quite that clear-cut though, due to **crossing over** (see page 96). This means the results might turn out like in this diagram of a fruit fly cross — the gametes **Sg** and **sG** (marked with a star) are **only** produced as a result of **crossing over**, so they occur much **less often** than you'd expect. In this example, it happens just **15%** of the time.

Inheritance

In **Mammals** Sex is Determined by the **X** and **Y** Chromosomes

The genetic information for your **gender** is carried on two **specific** chromosomes:

1) In mammals, **females** have **two X** chromosomes and **males** have **one X** and **one Y**. The probability of having male or female offspring at each pregnancy is **50%**.
2) The Y chromosome is **smaller** than the X chromosome and carries **fewer genes**. So most genes carried on the sex chromosomes are only carried on the X chromosome. These genes are sex-linked. Males only have **one copy** of the genes on the X chromosome. This makes them more likely than females to show **recessive phenotypes**.
3) Genetic disorders inherited this way include **colour-blindness** and **haemophilia**. The pattern of inheritance can show that the characteristic is **sex-linked**. In the example below, females would need **two copies** of the recessive allele to be colour blind, while males only need one copy. This means colour blindness is **much rarer** in **women** than **men**.

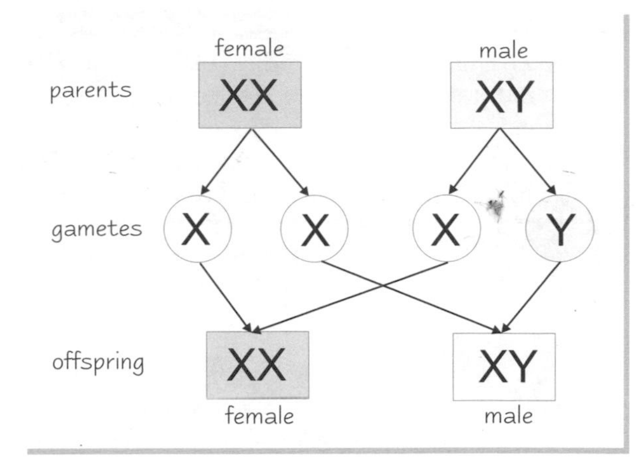

A carrier is a person carrying an allele which is not expressed in the phenotype, but which can be passed on.

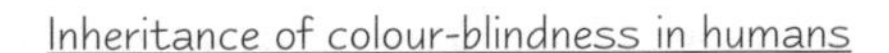

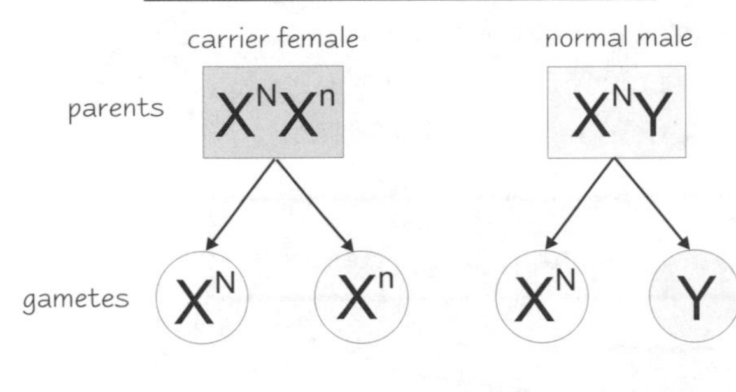

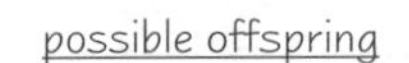

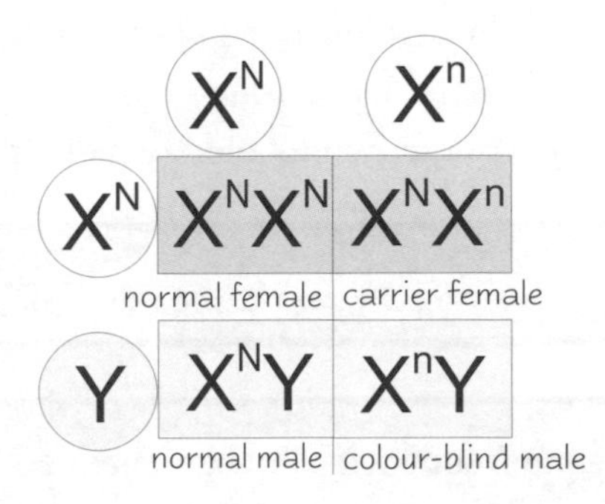

N = allele for normal eyesight
n = allele for colour-blindness

A carrier is a person carrying an allele which is not expressed in the phenotype, but which can be passed on.

Practice Questions

Q1 What is the expected ratio for dihybrid inheritance?

Q2 Which sex chromosome is smaller and carries fewer genes?

Q3 Why is red-green colour-blindness much more common in males than in females?

Exam Questions

Q1 Draw a genetic diagram to show the expected results of a cross between a normal-winged grey-bodied fruit fly (genotype NnGg) and a normal-winged ebony-bodied fruit fly (Nngg). [4 marks]

Q2 The recessive allele for haemophilia is carried on the X chromosome.
Explain why you would expect haemophilia to be more common in males than females. [5 marks]

Pedigree charts aren't just for dogs... they're for royals too...

You can use pedigree charts to track the inheritance of sex-linked conditions like haemophilia. The classic example is the inheritance of haemophilia from Queen Victoria, who carried the allele and spread it through the royal families of Europe.

Natural Selection and Evolution

Darwin's book on his theory of evolution 'On The Origin of Species' is probably the most important biology book that's ever been published. Apart from this one of course.

Darwin wrote his **Theory of Evolution** in **1859**

In **1831** Darwin was invited to join the ship **HMS Beagle** on a map-making trip around the world. Darwin was a keen **naturalist** and on the voyage he collected lots of **data** and many **samples** of **plants** and **animals**. When he returned he spent twenty years studying his data and eventually he came up with the **theory of evolution** which was published in **1859** in a book called **'On The Origin of Species by means of Natural Selection'**. Darwin's theory was based on five main assumptions:

1) **More individuals** are **produced** than can **survive**.
2) There is a **struggle** for **existence**.
3) Individuals within a species show **variation**.
4) Those with **advantageous features** have a greater chance of **survival**.
5) Those individuals who **survive** produce **similar offspring**.

Darwin used the term **'natural selection'** to describe the way that individuals with variations that help them survive in their habitat have advantages which make it more likely that they'll be able to pass on their genes. Darwin believed that natural selection **caused evolution**.

The **Latest** evolutionary theories still include **Natural Selection**

Obviously, there has been a lot more scientific research into natural selection done since Darwin published his book. The latest developments in genetics have been incorporated into Darwin's theory to update it:

1) There are changes in the **genetic composition** (gene frequencies) of a population from one generation to the next.
2) These changes are brought about by **mutation**, **genetic drift**, **gene flow** and **natural selection**:

Mutation

The **mutation** of genes can produce different **allele** and **phenotype frequencies** (see pages 98-99).

Genetic Drift

This is the **alteration** of **gene frequencies** through **chance**. For example, if two **heterozygous** individuals breed, their offspring may not be produced in an exact **Mendelian** ratio — the gene frequencies in populations change over time.

Gene Flow

This happens when new genes enter or leave a population by **migration**.

Natural Selection

1) As **conditions** change, or organisms **move** into a new environment, the organisms that are better **adapted** to the new conditions, because of the **alleles** they carry, will **survive**.
2) **Variation** between **isolated populations** increases as **gene pools diverge**.
3) Changes occur in **allele**, **genotype** and **phenotype** frequencies.
4) Eventually a **new species** will evolve.

Limiting Factors affect Survival and Reproduction Rates

There's more about the type of things that act as limiting factors on pages 88 and 89.

All organisms tend to **overproduce** — this inevitably brings about **intraspecific competition** for resources. **Limiting factors** like **parasites** put **selection pressures** on the organisms. This is when natural selection comes into play — the individuals that are best adapted to the conditions, because of the genes that they carry, are more likely to survive and reproduce. This process changes the **allele frequencies** in the population. For example, there would be a **greater number** of **individuals** with **resistance** to certain **parasites** in the population.

Natural Selection and Evolution

There are *Different Types* of *Selection*

Selection can have **different effects** on a population:

1) **Stabilising** selection is where individuals with traits towards the **middle** of the range are more likely to survive and reproduce. It's the **commonest** type, which occurs when the environment is **not** changing. It helps to keep the population **stable**.
2) **Directional** selection is where individuals of **one extreme type** are more likely to survive and reproduce. This happens when the environment changes, and it causes corresponding **genetic change** in the population.
3) **Disruptive** selection is where **two different extreme types** are selected for, perhaps because they live in **two different habitats**. This leads to **two distinct types** developing, and eventually these may become **different species**.

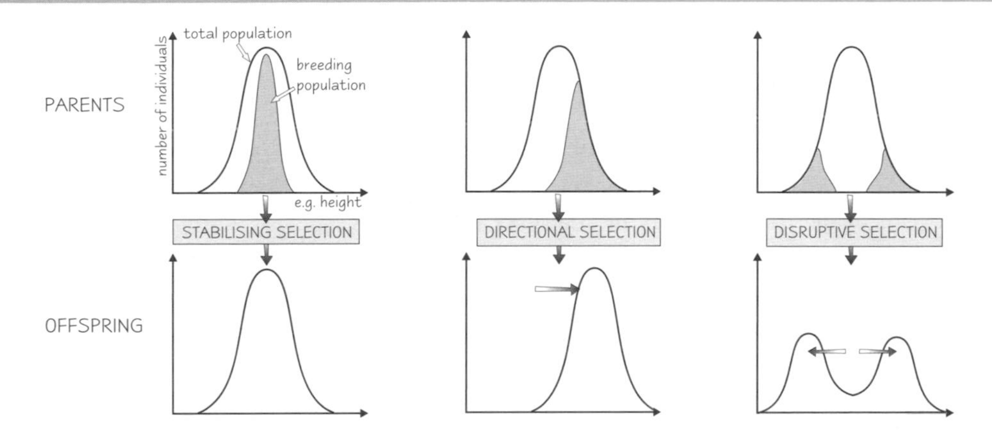

Practice Questions

Q1 How has modern biology updated Darwin's theory of evolution?

Q2 What does genetic drift mean?

Q3 What is gene flow?

Q4 Write a paragraph explaining natural selection in your own words.

Q5 Name the three different types of selection, and explain what effect each one has on allele frequencies.

Exam Question

Q1 In 1976-1977, a severe drought struck the Galapagos islands. No rain fell for over a year. During the drought a number of plant species died out. Some others did not produce seeds, causing a food shortage for the seed-eating ground finch *Geospiza fortis*.

The seeds of one plant species that survived the drought were stored in large, tough fruits. Only *Geospiza fortis* individuals with a beak depth greater than 10.5 mm were able to feed from this plant.

A biologist conducted a survey of the finches and recorded the mean beak depths and lengths of birds that survived and birds that died during the drought. The results of the survey are recorded in this table:

	Beak Length	Beak Depth
Surviving birds	11.07	9.96
Dead birds	10.68	9.42

As the population of *G. fortis* recovered after the drought, the mean beak depth of the population was greater than before (an increase of 4–5%). Explain this change with reference to evolutionary change. [6 marks]

The best type of selection comes in a box at Christmas...

Natural selection is the most important thing on this page so make sure that you read that bit thoroughly. You need to be able to describe how natural selection happens and what impact it has on future generations of the species. Also, check that you know how genetics has been incorporated into modern thinking on evolution.

Speciation

Speciation is all about how new species appear — there are two kinds of speciation that you need to know about — allopatric speciation and sympatric speciation.

Speciation is the Development of a New Species

A **species** is defined as a group of organisms that can **reproduce** and produce **fertile young**.
Every true species we know of has been named using the **binomial system** (see page 83). When new species are discovered they are also classified using this system.

Sometimes two individuals from different species can breed and produce offspring. These **hybrid** offspring aren't a new species because they're **infertile**. For example, **lions** and **tigers** have bred together in zoos to produce **tigons** and **ligers** but they aren't new species because they **can't** produce offspring.

Speciation (development of a new species) happens when **populations** of the **same species** become **isolated**.
Local populations of a species are called **demes** (be careful with this — demes aren't always isolated).

Geographical Isolation causes Allopatric Speciation

1) Geographical isolation happens when a **physical** barrier **divides** a population of a species.
2) **Floods**, **volcanic eruptions** and **earthquakes** can all lead to physical barriers that cause some individuals to become **isolated** from the main population.
3) **Conditions** on either side of the barrier will be slightly **different**. For example, there might be a different **climate** on either side of the barrier.
4) Environmental conditions like this put **pressure** on the organisms, forcing them to **adapt** — the **natural selection processes** differ in each isolated group.
5) **Mutations** will take place **independently** in each population and, over a **long** period of time, the gene pools will **diverge** and the **allele frequencies** will **change**.
6) Eventually, individuals from different populations will have changed so much that they won't be able to breed with one another to produce **fertile** offspring — they'll have become **two separate species**.

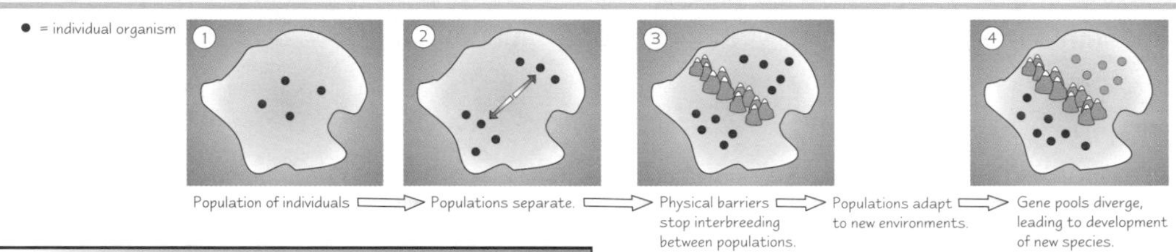

Isolation doesn't have to be Physical

Reproductive isolation happens when something **prevents** some members of a population breeding with each other.

There are **several causes** of reproductive isolation:

1) **Seasonal isolation** — where a mutation or genetic drift means that some individuals of the same species have different **flowering** or **mating** seasons, or become **sexually active** at different times of the year.
2) **Mechanical isolation** — where mutations cause changes in genitalia which prevent successful mating.
3) **Behavioural isolation** where a group of individuals develop **courtship rituals** which are **not attractive** to the main population of a species.
4) **Gametic isolation** — where mutations mean that male and female **gametes** from different populations of the same species are **not** able to create new individuals — so the individuals can mate, but fertilisation fails or the fetus is aborted.

If two populations have become so different that they can't breed then a **new species** will have been created — this is called **sympatric speciation**.

Plant Speciation can occur through Polyploidy

Sometimes cell division doesn't happen when gametes form, and some gametes end up being **diploid** rather than **haploid**. If these gametes fuse with other gametes you end up with individuals that have one or more **extra sets of chromosomes** — that's **polyploidy**. Sometimes, the chromosome set doubles **after fertilisation** — the chromosomes replicate as they would before mitosis, but then the cell doesn't divide. This **post-fertilisation polyploidy** becomes important if two closely related species are crossed. The offspring would be **sterile**, because the chromosomes would be **non-homologous** and so couldn't pair up during meiosis. But if the diploid number **doubles**, each chromosome **will** have a homologous one to pair with and meiosis **can** happen. This is thought to have happened to produce the modern **wheat** plant.

Speciation

Darwin's Finches are a good example of Allopatric Speciation

Darwin studied **finches** that live on the Galápagos Islands, a small group of islands 1000 km west of Ecuador, to develop his theory of evolution. He based his theory on his observations:

1) On the Galápagos islands, there are **fourteen** species of **finch** belonging to **four genera**.
2) Each species of finch inhabits a different ecological niche (see p.88) on the islands and some are only found on one island.
3) The main difference between the finches is the **shape** and **size** of their **beaks**. The birds feed on a variety of different foods from grubs to hard-shelled seeds — each finch has a beak suited to the food it eats.

main food	fruits	insects	insects	cacti	seeds	seeds
feeding adaptation	parrot-like beak	grasping beak	uses cactus spines	large crushing beak	pointed crushing beak	large crushing beak

Despite these differences, Darwin thought that all the finches had a **common ancestor**.
Since then more research has been done which has proved that **geographical isolation** did cause **speciation** on the Galápagos islands. Finches are small birds and it's unusual for them to fly over water, so once a population gets onto an island (perhaps because they were blown off course by a storm) they are effectively **isolated** from the finches on other islands. The differing environmental conditions on each island put **selection pressures** on the birds — and the birds gradually became **adapted** by natural selection to the conditions on the different islands.

Convergent Evolution is when Unrelated Species have Similar Features

Convergent evolution happens when **unrelated** species have **evolved** so that they look very **similar**. For example, **sharks** and **dolphins** look pretty similar and swim in a similar way but they're totally different species — sharks are cartilaginous **fish** and dolphins are **mammals**. They have different evolutionary roots but they have developed similar skeletal structures to make them well **adapted** for swimming.

Practice Questions

Q1 Define the term 'species'.
Q2 What is a hybrid? Give an example.
Q3 What is the difference between allopatric and sympatric speciation?
Q4 Name four causes of sympatric speciation.
Q5 What was Darwin researching when he proposed his theory of evolution?

Exam Question

Q1 Charles Darwin studied different species of finch on the Galápagos Islands.
a) Describe Darwin's observations. [3 marks]
b) Give an explanation of how Darwin believed the different species developed. [4 marks]

I wish there were biology field trips to the Galápagos Islands...

It's easy to learn the basics of these pages — what a species is and how new ones develop. Then it's just a matter of learning the detail and the correct words for everything. It's important that you know words like 'sympatric' and 'allopatric speciation' because they might be used in the exam questions and you'll be stuck if you forget what they mean.

Genetic Engineering

On these pages, you will learn how to splice rabbit DNA with that of an onion to make a new and exciting pet.

Genetic Engineering has Loads of Important Uses

Genetic engineering (or **genetic manipulation**) is when DNA is removed from one organism and joined to the DNA of another. DNA treated in this way is called **recombinant DNA**.

There are already many uses of genetic engineering, and there are potentially loads more. **Herbicide-resistant** crop plants are now created in this way. **Bacteria** are engineered to produce useful **proteins** such as **human insulin**. It is even possible to over-ride defective alleles in **embryos** by genetic engineering. (This only works for **recessive** conditions — the normal functioning allele is added to cells, with the faulty allele still in place.)

Useful Genes can be Isolated with Restriction Endonuclease Enzymes

The first stage in genetic engineering is removing the useful DNA from the **donor** cell:

1) Any proteins attached to the DNA are digested by **peptidase** enzymes.
2) The useful gene is cut out of the DNA using **restriction endonuclease** enzymes. These enzymes are normally found in bacteria. Their usual job is to destroy the DNA or RNA of invading viruses by chopping it into small pieces.
3) Restriction endonucleases leave the fragment of DNA with a small tail of unpaired nucleotide bases at each end. These tails are called **sticky ends**.

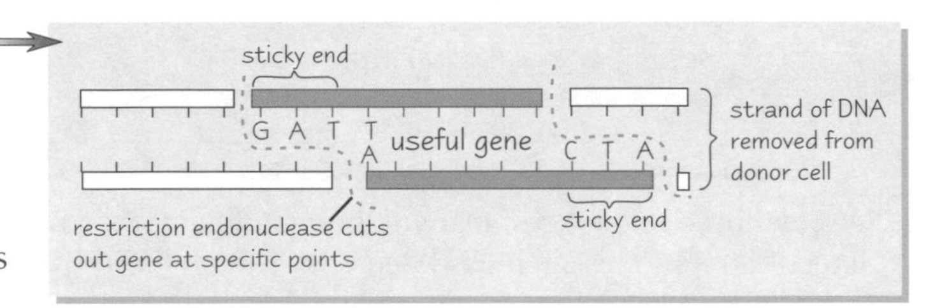

Vectors Carry the Useful Gene

1) The useful gene is now joined to a **vector** (the DNA that carries the gene into the bacterium). The main vectors used are **plasmids** — these are small, **circular molecules** of DNA found in **bacteria**. They're useful because they **replicate** easily and can be put into other bacteria without harming them. One plasmid often used in genetic engineering contains genes for **resistance** to the antibiotics **ampicillin** and **tetracycline** — the reason for this is explained on page 109. DNA from a **bacteriophage virus** (see page 28) can also be used as a vector.

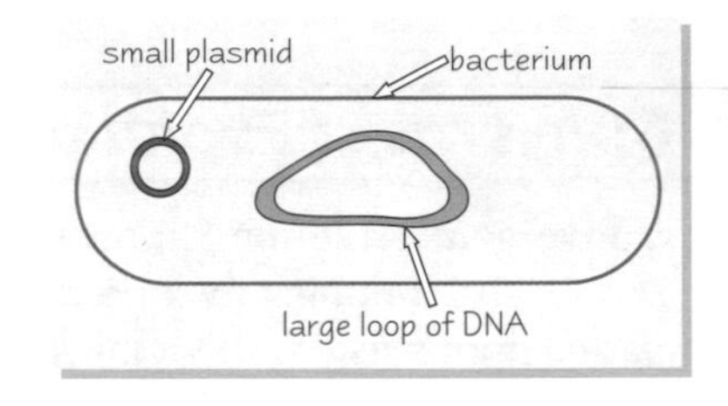

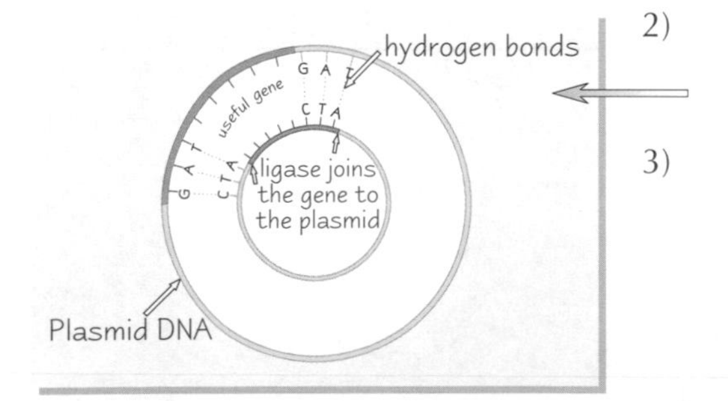

2) The same **restriction endonuclease** that was used to remove the useful gene from the **donor** is used to cut open the plasmid. The **sticky ends** of the open plasmid are **complementary** to those of the useful gene.
3) The useful gene is **joined** to the **plasmid vector DNA** using those clever **sticky ends**. This is known as **splicing**. The complementary bases pair by **hydrogen bonding** and the DNA is then joined by the enzyme **DNA ligase**. This process of joining is called **ligation** and the new combination of bases in the DNA (plasmid DNA + useful gene) is called **recombinant DNA**.

Recombinant DNA is put into the Host Bacterium

The plasmid containing the useful gene now has to be put into a **bacterium** so it can be **replicated** and produce its useful **protein**. A harmless strain of the bacterium *Escherichia coli* is often used.

1) The bacteria and the recombinant plasmids are added to **ice cold calcium chloride solution**.
2) The mixture is heated rapidly to 42 °C for 50 seconds. This **heat shock** allows the plasmids to get through the bacterial cell wall and membrane and into the cytoplasm.
3) The mixture is quickly **cooled** on ice.
4) Bacteria that have successfully taken up the plasmids are called **transformed bacteria**.

Genetic Engineering

Transformed Bacteria *can be Located Using* ***Marker Genes***

Only a few of the bacteria will have taken up the vectors, so the next step is to find out which ones have.
Antibiotic marker genes can show which bacteria have been **transformed**.

1) The useful gene is inserted into the plasmid, in the **middle** of the gene for **tetracycline resistance** (see page 108). This means that the gene for tetracycline resistance **can't work** any more, but the gene for **ampicillin resistance** still functions normally.
2) After transformation, the bacteria are **cultured** on an agar plate called the **master plate**. The master plate contains **ampicillin**, so any bacteria that haven't taken up any plasmids will be **killed**.
3) Next, a **sterile velvet pad** is pressed lightly onto the surface of the master plate. This picks up a few from each colony of bacteria. The pad is pressed onto an agar plate (the **replica plate**) containing **tetracycline**. Some bacteria from each colony are transferred to the replica plate.
4) Bacteria containing the **transformed plasmid** will be **killed**, because their gene for tetracycline resistance has been **disrupted** by the inserted gene (the corresponding colony on the master plate can still be used as a source of these bacteria). Bacteria that **survive** on the replica plate have only taken up **non-recombinant plasmids**.

Bacteria can be Cultured on an ***Industrial Scale***

1) The useful gene usually needs a '**kick-start**' before it will work, so a **promoter gene** is often included when the donor and antibiotic resistance genes are **spliced** into the plasmid. The promoter gene **switches on** the useful gene so it will begin producing the **protein**.
2) The bacteria are cultured in optimum conditions on an **industrial scale**. As they reproduce **asexually** by binary fission, they make **identical copies** of the useful gene.
3) The gene causes the bacteria to produce the useful **protein**, which the bacteria can't use, so it builds up and can be **extracted** and **purified**.

Reverse Transcriptase *makes DNA from Messenger RNA*

Finding a **single** copy of a useful gene is much more difficult than finding **lots** of mRNA molecules.
This is why the enzyme **reverse transcriptase** is very useful — it makes DNA from mRNA (the opposite of normal **transcription** — which you covered at AS), so it can be used to make certain useful genes.

1) Genetic engineers can **isolate** the right mRNA, then use this enzyme to make the **complementary DNA** (cDNA).
2) The DNA made by the reverse transcriptase is **single-stranded**, so the enzyme **DNA polymerase** is used to make the full double-stranded version.

Practice Questions

Q1 What is a sticky end?

Q2 Give two examples of vectors used in genetic engineering.

Q3 How does a replica plate differ from a master plate?

Exam Questions

Q1 *Aequorea victoria* is a species of jellyfish that fluoresces (glows). The gene that codes for the protein responsible for this fluorescence has been discovered. It is possible to insert this gene into the bacterium *Escherichia coli*.

a) Briefly describe the stages involved in the transformation of the bacterium. [9 marks]

b) Suggest a use for this gene and explain your answer. [2 marks]

Q2 Explain why reverse transcriptase is so useful in genetic engineering. [6 marks]

OK, so I lied about the onion bunny (or bunnion)...

It was a cheap trick to get your attention. But seriously, genetic engineering is really important — it means that people with genetic diseases now have a much better chance of successful treatment. There are some dodgy ethical issues associated with it, like designer babies and stray genes 'escaping' — but genetic engineering is still a pretty exciting breakthrough.

Commercial Use of Genetic Engineering

Genetic engineering benefits all kinds of people, from diabetics to vegetarians. But some people worry that it could lead to a genetic underclass, new diseases, superweeds, the end of the world...

Microorganisms can Produce Proteins on an *Industrial Scale*

Microorganisms include things like **bacteria**, **yeasts** (fungi) and **protoctists**. They're great for mass-production of genetically-engineered products. Here are seven reasons why:

1) Microorganisms have a very **rapid growth rate**. In optimum conditions they can double their biomass in 20 minutes.
2) There's a huge variety of **substrates** they can use as **food**. Some even feed on oil and plastic — so cheap waste can be used up too.
3) Their **genome** (the total collection of genes in an organism) is easily **manipulated**.
4) They can be easily **screened** (tested) to see if they have desirable characteristics (e.g. fast growth rate).
5) They can be grown in **fermenters** in any **climate**, unlike multicellular organisms which often need conditions that are difficult to provide.
6) **Fermenters** take up relatively little **space**.
7) The useful products can be **isolated** and **purified** quite easily.

Insulin and *Human Growth Hormone* are Produced for Use in *Medicine*

This flow chart shows how insulin is produced on an industrial scale. See pages 16-17 for more on insulin.

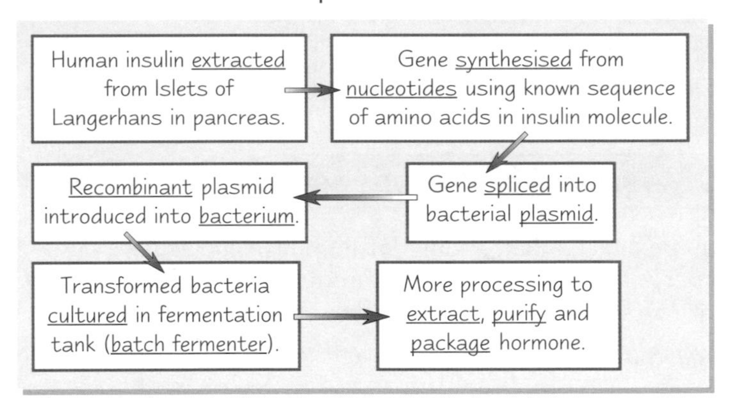

See pages 34-35 for more on batch fermenters.

Human growth hormone is also produced in this way. Insulin used to be extracted from the pancreas of dead pigs and human growth hormone from the pituitary of human corpses — not nice.

Genetic engineering, batch fermentation and downstream processing ensure that **pure**, **human** proteins are produced in **large** quantities, relatively **inexpensively**.

Chymosin is a Genetically Engineered Alternative to *Calf Rennin*

Rennin is an enzyme used to **coagulate** the protein in milk to make **cheese**. It's found in the **abomasum** (fourth stomach) of calves.

The gene for rennin production is inserted into **yeast cells**. Cultures of yeast in fermenters can then produce large amounts of the enzyme, which is now called **chymosin** (but it's the same as rennin).

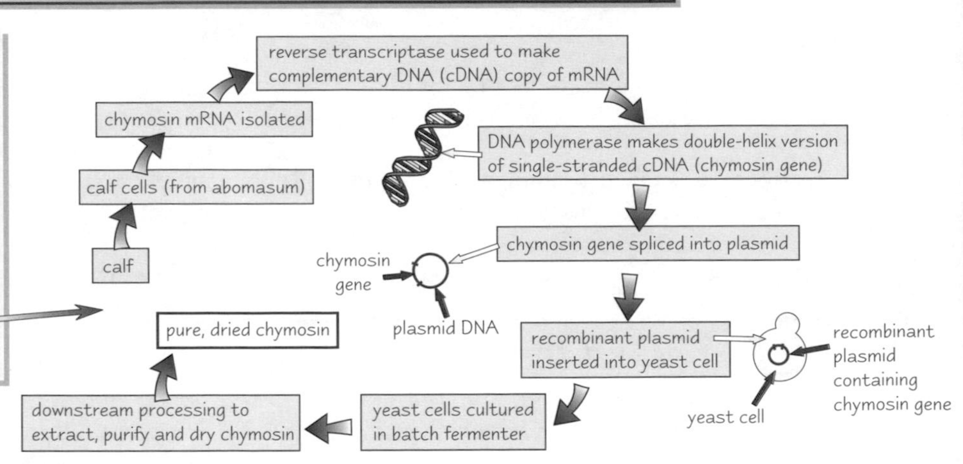

Commercial Use of Genetic Engineering

Crop Plants can be *Improved* by Genetic Engineering

Genes from unrelated species can be inserted into **crop plants** to give them desirable characteristics.

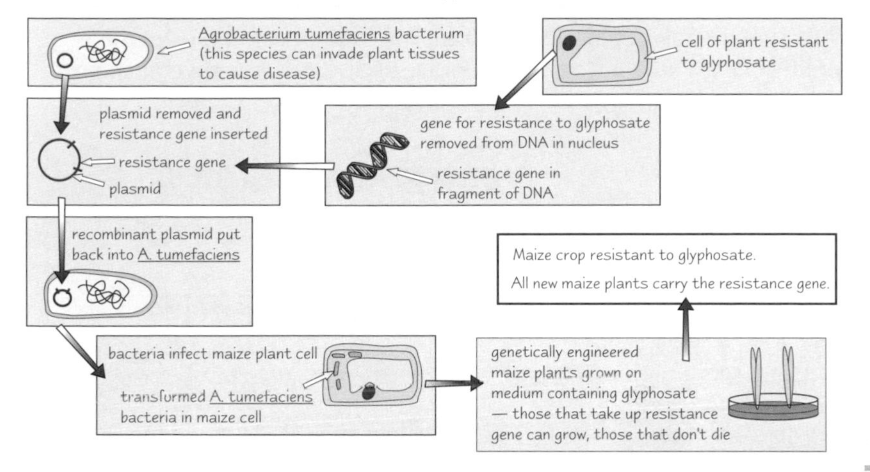

- A gene giving resistance to the herbicide **glyphosate** can be inserted into crop plants such as **maize**. When the crop is sprayed with herbicide, the weeds die but the maize is unaffected.
- Scientists have inserted genes into broccoli to increase the amount of **glucosinates** — anti-cancer chemicals.
- **Cotton plants** have been genetically modified with a gene from the bacterium *Bacillus thuringiensis* that produces an **insecticide**. Any insects that try to eat the plants are killed by the insecticide.

Many People are *Worried* about Genetic Engineering

There are lots of social and economic **benefits** of genetic engineering, but there are **hazards** too. Some people think all kinds of **genetic manipulation** are **unethical**.

Benefits	Hazards
Medically important proteins no longer need to be extracted from organs, but can now be produced by transformed bacteria (e.g. erythropoietin is a protein essential for regulation of red blood cell production).	No control of recombinant DNA once released into the environment (e.g. genes 'escaping' in pollen grains).
Hormones etc. can be tailor-made to exact requirements (e.g. diabetics now use human, not pig, insulin).	Viruses and bacteria containing recombinant DNA could mutate and become pathogenic, causing new diseases.
Crops can be improved and made herbicide-resistant (e.g. glyphosate-resistant maize).	Genetically modified organisms sometimes exhibit unpredicted characteristics.
Can be used to produce vegetarian products. E.g. the enzyme used in cheese-making can be mass-produced by genetically modified yeast, rather than using animal enzymes.	Weeds and pest insects could gain rogue genes from engineered crops and become resistant to chemicals designed to kill them.
Genetic diseases can be treated by adding the functioning allele. (Cystic fibrosis is already treated this way.)	Risk of creating 'GM monsters' (new animals produced by combining genes from different species).

Bioethics concerns the **life sciences** and their impact on **society**. Genetic engineering raises some important **bioethical** issues:

- Some people think it's wrong to genetically engineer animals purely for **human benefit**, especially if the animal **suffers**.
- Those who can afford it might decide which characteristics they wish their children to have (**designer babies**), creating a 'genetic underclass'.
- The **evolutionary consequences** of genetic engineering are unknown.
- There are **religious concerns** about 'playing God'.
- **Xenotransplantation** (transplanting animal organs engineered to resist rejection) can save lives, but there are **religious** implications. For example, many Muslims would find it unthinkable to receive a transplant from a pig.

Practice Questions

Q1 List four reasons why microorganisms are useful in mass production of genetically engineered products.

Q2 Where is rennin normally found? What is it used for commercially?

Q3 Give two benefits and two hazards of genetic engineering.

Exam Questions

Q1 Describe how a named GM human hormone is produced. [5 marks]

Q2 Describe three benefits and three drawbacks of creating genetically modified herbicide-resistant crops. [6 marks]

Designer babies — mine's Gucci, daahling...

Here are some websites you might find interesting: www.bioethics-today.org (Bioethics Today – a bioethics resource for the UK) and www.newint.org/issue215/monsters.htm ('Monsters of the Brave New World' – an article on GM monsters).

The PCR and Genetic Fingerprinting

These are good pages for those of you considering a career in crime. It's a bit more tricky these days because your own body is working against you — the tiniest hair or flake of skin it drops is enough to place you at the scene of the crime.

The Polymerase Chain Reaction (PCR) Creates Millions of Copies of DNA

Some samples of DNA are too small to analyse. The **Polymerase Chain Reaction** makes millions of copies of the smallest sample of DNA in a few hours. This **amplifies** DNA, so analysis can be done on it. PCR has **several stages**:

1) The DNA sample is **heated** to **95°C**. This breaks the hydrogen bonds between the bases on each strand. But it **doesn't** break the bonds between the ribose of one nucleotide and the phosphate on the next — so the DNA molecule is broken into **separate strands** but doesn't completely fall apart.
2) **Primers** (short pieces of DNA) are attached to tiny bits of both strands of the DNA — these will tell the **enzyme** where to **start copying** later in the process. They also stop the two DNA strands from joining together again.

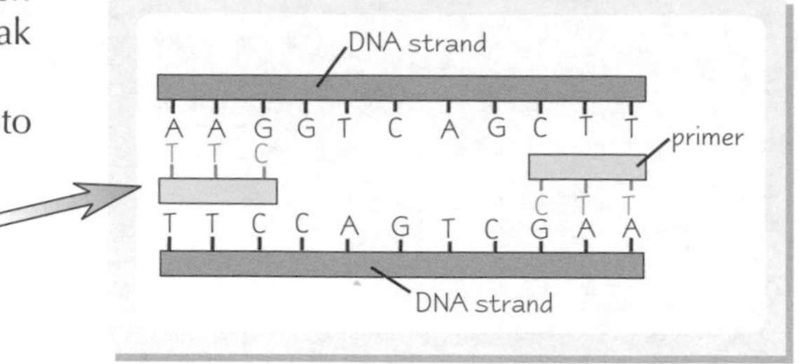

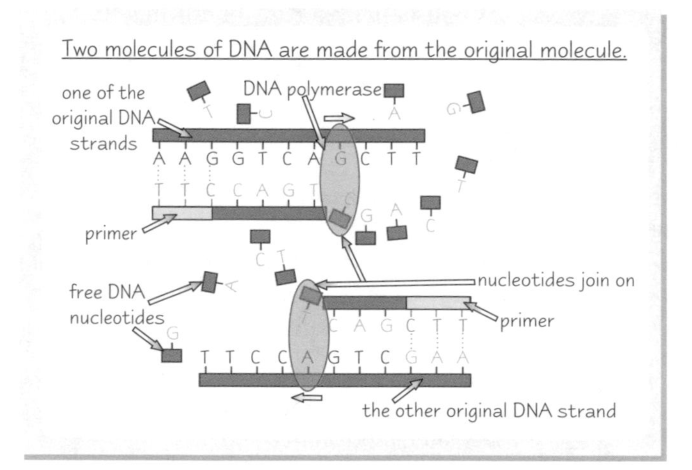

3) The DNA and primer mixture is **cooled** to **40°C** so that the primers can **fully bind on** to the DNA.
4) Free **DNA nucleotides** and the enzyme **DNA polymerase** are added to the reaction mixture. The mixture is heated to **70°C**. Each of the original DNA strands is used as a **template**. Free DNA nucleotides pair with their complementary bases on the template strands. The DNA polymerase attaches the new nucleotides together into a strand, starting at the primers.
5) The cycle starts again, using **both** molecules of DNA. Each cycle **doubles** the amount of DNA.

You can Identify People from their DNA by Cutting it into Fragments

It's possible to **identify a person** from a sample of their DNA, if the sample is big enough. This is done by using **enzymes** to cut the DNA up into **fragments**, then looking at the **pattern** of fragments, which is **different** for everyone. This is called a person's **genetic fingerprint**.

1) To **cut up** the DNA into DNA fragments you add specific **restriction endonuclease** enzymes to the DNA sample — each one **cuts** the DNA every time a **specific base sequence** occurs. The **location** of these base sequences on the DNA **varies** between everyone, so the number and length of DNA fragments will be different for everyone.
2) Next you use the process of **electrophoresis** to separate out the DNA fragments by size:

How Electrophoresis Works:

1) The DNA fragments are put into **wells** in a slab of **gel**. The gel is covered in a **buffer solution** that **conducts electricity**.
2) An **electrical current** is passed through the gel. DNA fragments are **negatively charged**, so they move towards the positive electrode. **Small** fragments move **faster** than large ones, so they **travel furthest** through the gel.
3) By the time the current is switched off, all the fragments of DNA are **well separated**.

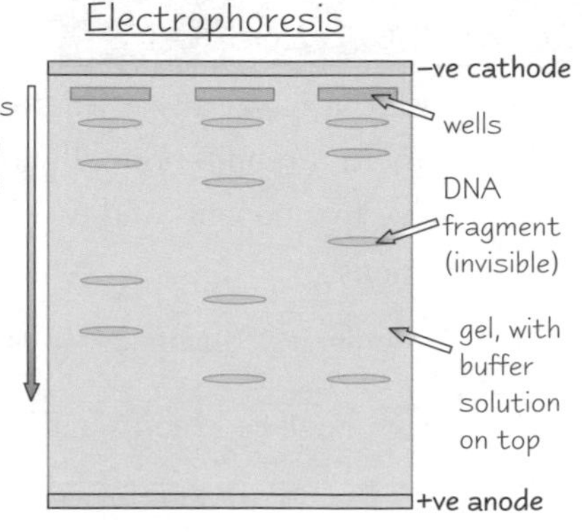

In electrophoresis the DNA fragments **aren't visible** to the eye — you have to do something else to them before you can **see their pattern**. Coincidentally, that's what the next page is all about...

The PCR and Genetic Fingerprinting

Gene Probes Make the **Invisible** 'Genetic Fingerprint' **Visible**

DNA fragments separated by electrophoresis are invisible.
A radioactive DNA probe (also called **gene probe**) is used to show them up:

1) A **nylon membrane** is placed over the electrophoresis gel and the DNA fragments **bind** to it.
2) The DNA fragments on the nylon membrane are **heated** to separate them into **single strands.**
3) **Radioactive gene probes** are then put onto the nylon membrane. (It's the **phosphorus** in the gene probes' sugar-phosphate backbones that's radioactive.) The probes are warmed and **incubated** for a while so that they'll attach to any bits of **complementary DNA** in the DNA fragments.

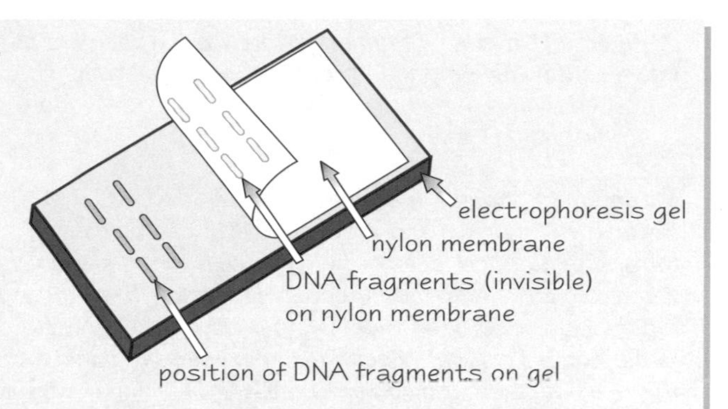

4) The nylon membrane is then put on top of unexposed **photographic film**. The film goes **dark** where the radioactive gene probes are **present**, which **reveals the position** of the **DNA fragments**.

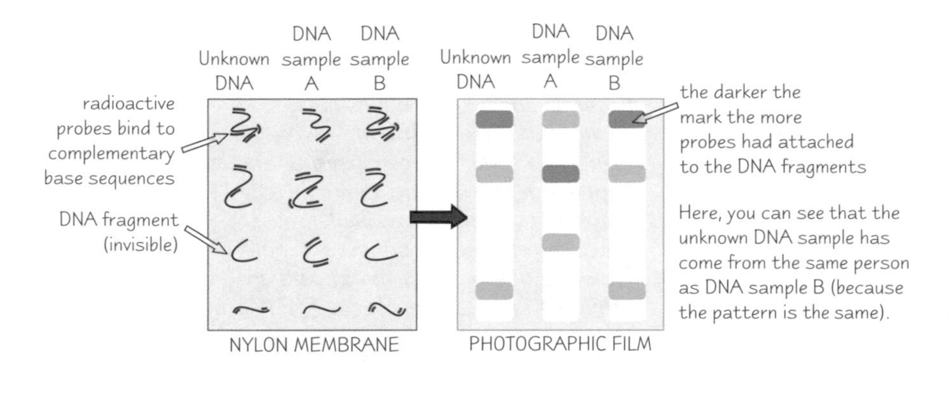

the darker the mark the more probes had attached to the DNA fragments

Here, you can see that the unknown DNA sample has come from the same person as DNA sample B (because the pattern is the same).

Genetic fingerprinting is incredibly useful. **Forensic investigations** use it to confirm the identity of suspects from blood, hair, skin, sweat or semen samples left at a crime scene, or to establish the identity of victims. **Medical investigations** use the same technique for **tissue typing, paternity tests** and **infection diagnosis.**

Practice Questions

Q1 Give the full name of the technique used to increase the amount of DNA from a very small sample.

Q2 Name the enzyme used in PCR.

Q3 What types of enzyme are added to DNA before gel electrophoresis?

Q4 Which part of a gene probe is radioactive?

Q5 State three uses of genetic fingerprinting in medical investigations.

Exam Question

Q1 Police have found incriminating DNA samples at the scene of a murder.
They have a suspect in mind and want to ascertain if the suspect is guilty.

a) Name the technique that the police could use to confirm the guilt of the suspect. [1 mark]

b) Explain how this technique would be carried out. [5 marks]

Hands up, punk — I've got a gene probe and I'm not afraid to use it...

These days, anyone who gets arrested by the police, for whatever reason, has a sample of their DNA taken and kept on file. It won't be long until most people's are recorded. Wherever you go, you're leaving little traces of your DNA behind. So one day, we'll just hoover up all the DNA at the scene of a crime and know immediately who was there. Scary thought.

Answers

Section 1 — Respiration

Page 3 — Energy and the Role of ATP

1 *Maximum of 2 marks available.*
ATP is a molecule made from adenosine diphosphate (ADP) and phosphate, using energy from reactions like those of respiration **[1 mark]**. *The energy stored in the chemical bond between the ADP and the phosphate can be released when it is needed by a cell, by breaking the ATP back down into ADP and phosphate* **[1 mark]**.

2 *Maximum of 3 marks available.*
Because ATP is small and water-soluble, it can be easily transported around a cell or between cells to the places where there is a demand for energy **[1 mark]**. *There it can be rapidly converted back into ADP to release the energy stored in the bonds* **[1 mark]**. *Because an enzyme is required for this reaction, there is little risk of the ATP breaking down into ADP and releasing its energy in the wrong place, wasting the energy* **[1 mark]**.

Page 5 — Glycolysis

1 *Maximum of 5 marks available.*
The 6-carbon glucose molecule is hydrolysed / split using water **[1 mark]**, *and phosphorylated using phosphate from 2 molecules of ATP* **[1 mark]** *to give 2 molecules of the 3-carbon molecule glycerate 3-phosphate* **[1 mark]**. *This is then oxidised by removing hydrogen ions* **[1 mark]** *to give 2 molecules of 3-carbon pyruvate* **[1 mark]**.

2 *Maximum of 4 marks available.*
The 3-carbon pyruvate is combined with coenzyme A **[1 mark]** *to form a 2-carbon molecule, acetyl coenzyme A* **[1 mark]**. *The extra carbon is released as carbon dioxide* **[1 mark]**. *The coenzyme NAD is converted into reduced NAD in this reaction by accepting hydrogen ions* **[1 mark]**.

You need to know how many carbons there are in each molecule. That decides the molecule's basic structure, and it's the most important change from molecule to molecule in these big chains you have to know about. Learn the names of the molecules by all means, but if you forget one in the exam and just put '3-carbon compound', I bet you get the marks.

Page 7 — Krebs Cycle and Electron Transport Chain

1 *Maximum of 14 marks available.*
1 mark can be awarded for any of the following points, even if the final answer is incorrect:
2 ATP are produced in glycolysis **[1 mark]**. *1 ATP is produced per turn of the Krebs cycle* **[1 mark]**, *which happens twice per molecule of glucose* **[1 mark]** *giving 2 ATP from the Krebs cycle per molecule of glucose* **[1 mark]**.
In the electron transport chain, 2.5 ATP are produced for every molecule of reduced NAD coenzyme made in the earlier stages of respiration **[1 mark]**, *and 1.5 ATP for every molecule of reduced FAD produced* **[1 mark]**.
2 reduced NAD are produced in glycolysis **[1 mark]**, *1 reduced NAD is produced in the link reaction* **[1 mark]** *and 3 in the Krebs cycle* **[1 mark]**, *but for every molecule of glucose, 2 molecules of pyruvate are made by glycolysis* **[1 mark]**, *so the link reaction and Krebs cycle happen twice per molecule of glucose* **[1 mark]**.
So in total, 8 molecules of reduced NAD are produced by the link reaction and the Krebs cycle **[1 mark]**. *Adding the 2 reduced NAD produced in glycolysis gives 10 molecules of reduced NAD* **[1 mark]**. *$10 \times 2.5 = 25$ ATP* **[1 mark]**.
1 molecule of reduced FAD is also produced per turn of the Krebs cycle **[1 mark]**, *giving 2 reduced FAD per glucose molecule* **[1 mark]**. *$2 \times 1.5 = 3$ATP* **[1 mark]**. *So in total, the electron transport chain produces $25 + 3 = 28$ ATP* **[1 mark]**. *Adding the ATP produced in glycolysis and in the Krebs cycle gives $28 + 2 + 2 = 32$ molecules of ATP in total* **[1 mark]**.

Page 9 — Anaerobic Respiration

1 a) *Maximum of 10 marks available from the following points:*
The two forms of anaerobic respiration are alcohol fermentation and lactate fermentation **[1 mark]**.
Both are ways of releasing energy without using oxygen **[1 mark]**, *and both take place in the cytoplasm rather than in mitochondria* **[1 mark]**. *Both produce 2 ATP per molecule of glucose* **[1 mark]**, *and both begin by using the process of glycolysis* **[1 mark]** *to convert glucose into 2 molecules of pyruvate* **[1 mark]**.
They differ in what happens next — in alcohol fermentation, carbon dioxide is removed from the pyruvate to give 2-carbon ethanal **[1 mark]**. *A molecule of reduced NAD from glycolysis is then oxidised back to NAD* **[1 mark]**, *and the hydrogen ions it gives up are accepted by the ethanal, making ethanol (alcohol)* **[1 mark]**. *In lactate fermentation, no carbon dioxide is given off* **[1 mark]** *— reduced NAD is used to supply the hydrogen ions needed to reduce the pyruvate to lactic acid/lactate* **[1 mark]**. *Alcohol fermentation happens in plants and some micro-organisms, and lactate fermentation happens in animals and some bacteria* **[1 mark]**.

b) *Maximum of 3 marks available.*
Aerobic respiration produces 32 molecules of ATP per molecule of glucose **[1 mark]**, *and anaerobic respiration produces only 2 molecules of ATP per molecule of glucose* **[1 mark]**. *So in terms of ATP production, aerobic respiration is 16 times more efficient than anaerobic respiration* **[1 mark]**.

c) *Maximum of 2 marks available.*
The muscles of an Olympic sprinter running a 100 m race are working so hard that oxygen is used up very rapidly **[1 mark]**. *If the muscles did not respire anaerobically, they wouldn't be able to release any energy at all because there is no oxygen available* **[1 mark]**.

Section 2 — Physiological Control

Page 11 — Homeostasis and Communication

1 a) *Maximum of 2 marks available.*
A change in a factor brings about a response that counteracts the change **[1 mark]** *so that the factor returns to a norm* **[1 mark]**.

b) *Maximum of 2 marks available from any of the following.*
Body temperature **[1 mark]**, *blood glucose concentration* **[1 mark]**, *water potential* **[1 mark]**. *Or other sensible answer.*

2 a) *Maximum of 3 marks available for any of the following points:*
Receptors of the nervous system can communicate with effectors by releasing a chemical / hormone that binds to them **[1 mark]**. *The chemical / hormone moves from the receptor to the effector by diffusion if the two cells are close together* **[1 mark]**, *or if they are far apart the chemical / hormone is released into the blood and transported via mass flow* **[1 mark]**. *The other main way that receptors can send a message to effectors is to trigger an electrical impulse in a nerve, which is then passed through the nervous system until it reaches the effector and triggers a response* **[1 mark]**.

Answers

b) Maximum of 2 marks available from the following:
Chemical communication is slower than nervous communication **[1 mark]**. Chemical communication produces a longer response than nervous communication **[1 mark]**. Chemical communication uses the endocrine system and nervous communication uses the nervous system **[1 mark]**. Chemical communication uses hormones and nervous communication uses nervous impulses to transmit information **[1 mark]**. The effects of nervous communication are localised and the effects of chemical communication can be widespread **[1 mark]**.

Page 13 — The Kidneys and Excretion

1 Maximum of 6 marks available from the following:
Protein in food is digested **[1 mark]** into amino acids **[1 mark]**. If a lot of protein is digested, excess amino acids will be deaminated / have amino groups removed **[1 mark]** in the liver **[1 mark]** to form ammonia **[1 mark]**. The ammonia reacts with carbon dioxide **[1 mark]** to form urea **[1 mark]**, and more urea is excreted in the urine by the kidneys **[1 mark]**.

It's important to say that only the excess amino acids are deaminated.

2 a) Maximum of 5 marks available.
Microvilli provide a large surface area of membrane **[1 mark]** so there is a large surface area available for substances to pass across, and more carrier proteins **[1 mark]**. There are lots of mitochondria **[1 mark]** which provide energy / ATP for reabsorption **[1 mark]** by active transport **[1 mark]**.

b) Maximum of 3 marks available.
More glucose passes into the blood, because it is actively reabsorbed **[1 mark]**, but there is no carrier protein for urea **[1 mark]**. Some urea passes into the blood by diffusion **[1 mark]** because it's a small molecule **[1 mark]**.

3 Maximum of 2 marks available.
Either blood pressure is high(er) **[1 mark]**; so proteins get filtered out of blood **[1 mark]**. **Or** the wall of the capillary / basement membrane / wall of renal capsule is damaged / more perforated **[1 mark]** so that larger molecules are being filtered out **[1 mark]**.

Page 15 — The Kidneys and Osmoregulation

1 Maximum of 10 marks available.
Strenuous exercise causes more sweating **[1 mark]** so more water is lost from the body **[1 mark]**. This increases the blood solute concentration / decreases blood water potential / makes blood water potential more negative **[1 mark]**. It also stimulates osmoreceptors **[1 mark]** in the hypothalamus **[1 mark]**, which stimulates the posterior pituitary gland **[1 mark]** to release **more** ADH **[1 mark]**.

The answer up to this point has explained the cause of the increase in level of ADH in the blood. After this, the answer explains the effect on the kidney.

ADH increases the permeability of the collecting ducts **[1 mark]** so more water is reabsorbed into the blood by osmosis **[1 mark]**. This means that less water is lost in the urine which prevents further dehydration **[1 mark]**.

Page 17 — Blood Glucose Control

1 Maximum of 10 marks available from the following.
Glucose is absorbed into the blood (from the gut) **[1 mark]** which makes the blood glucose concentration increase **[1 mark]**. This stimulates the beta cells **[1 mark]** of the islets of Langerhans **[1 mark]** to secrete insulin **[1 mark]**. Insulin is released into the bloodstream **[1 mark]** and binds to receptors **[1 mark]** on liver cells **[1 mark]**. This increases the permeability of the liver cells to glucose **[1 mark]**, converting glucose into glycogen / stimulating glycogenesis **[1 mark]**. This reduces blood glucose concentration **[1 mark]**.

2 Maximum of 5 marks available.
Exercise uses up glucose (by respiration) **[1 mark]** which lowers blood glucose concentration **[1 mark]** and stimulates the alpha cells **[1 mark]** of the islets of Langerhans **[1 mark]** to secrete glucagon **[1 mark]**.

Page 19 — Chemical Communication

1 It takes time for the blood stream to carry them (from gland to target) **[1 mark]**.

2 Maximum of 6 marks available.
The hormone diffuses into the blood capillaries **[1 mark]**. The hormone is circulated around the body by mass flow **[1 mark]**. The hormone diffuses out of the blood and binds to cell-surface receptors **[1 mark]** of the target cells **[1 mark]** with complementary shaped binding sites **[1 mark]**. This brings about a response in the target cells **[1 mark]**.

Page 21 — Nervous Communication

1 a) Stimulus **[1 mark]**.

b) Maximum of 3 marks available.
A stimulus causes sodium channels in the neurone cell membrane to open **[1 mark]**. Sodium ions **diffuse** into the cell **[1 mark]**, so the membrane becomes depolarised / more positive inside **[1 mark]**.

c) Maximum of 2 marks available, from any of the following.
The membrane was in the refractory period **[1 mark]** and so the sodium channels were inactive / recovering / couldn't be opened **[1 mark]**. Alternatively, the stimulus could have been lower than threshold level **[1 mark]**.

2 Maximum of 5 marks available, for any 5 of the following:
Transmission of action potentials will be slower **[1 mark]**. Myelin insulates the axon / has high electrical resistance **[1 mark]** and there are gaps / nodes of Ranvier between sheaths **[1 mark]** where depolarisation happens /sodium channels are concentrated **[1 mark]**. So in an intact myelinated axon, saltatory transmission occurs / action potentials jump from node to node **[1 mark]**. This can't happen if the myelin sheath is damaged / more membrane is exposed **[1 mark]**.

Don't panic if a question mentions something you haven't learned about. You might not know anything about multiple sclerosis, and that's fine because you're not supposed to. All you need to know about to get full marks here is the structure of neurones.

Answers

Page 23 — Synapses

1 *Maximum of 8 marks available, for any 8 of the following:*
*Arrival of the action potential causes calcium channels (in the presynaptic membrane) to open **[1 mark]** which makes calcium ions diffuse into the bouton / synaptic knob **[1 mark]**. This stimulates the vesicles in the bouton to fuse with the presynaptic membrane **[1 mark]** and release the neurotransmitter **[1 mark]** by exocytosis **[1 mark]**.*
*The neurotransmitter diffuses across the gap / cleft **[1 mark]** and binds to receptors on the postsynaptic membrane **[1 mark]**. This stimulates opening of sodium channels (on the postsynaptic membrane) **[1 mark]** so sodium ions diffuse into the cell **[1 mark]**. The membrane then becomes depolarised and an action potential occurs **[1 mark]**.*

2 *Maximum of 4 marks available.*
*Vesicles (containing neurotransmitter) are only found in the presynaptic neurone **[1 mark]**, so exocytosis / release / secretion (of neurotransmitter) can only happen from here **[1 mark]**. The receptors (for the neurotransmitter) are only found on the postsynaptic membrane **[1 mark]** so only this membrane can be stimulated by the neurotransmitter / neurotransmitter can only bind here **[1 mark]**.*

Page 25 — The Central Nervous System

1 *Maximum of 4 marks available.*
*A sensory neurone enters the spinal cord **[1 mark]**.*
*It connects directly to a motor neurone **[1 mark]**. There is one synapse **[1 mark]**. An example is the knee jerk reflex **[1 mark]**.*

2 *Maximum of 4 marks available.*
*The hypothalamus sends action potentials / nerve impulses to the pituitary gland **[1 mark]**. The signal / chemicals from the hypothalamus pass to the pituitary gland **[1 mark]**.*
*The hypothalamus affects the activity of the pituitary gland **[1 mark]**, so it allows the nervous system to control the endocrine system **[1 mark]**.*

Section 3 — Microbiology and Biotechnology

Page 27 — Bacteria

1 *Maximum of 2 marks available.*
*Salmonella bacteria contain endotoxins **[1 mark]**, and their release prevents absorption of water in the large intestine, leading to diarrhoea **[1 mark]**.*

2 a) *Maximum of 2 marks available.*
*The layer of polysaccharide might prevent desiccation / drying out **[1 mark]** and attack from antibodies / phagocytes / white blood cells **[1 mark]**.*

b) *Maximum of 3 marks available.*
*The diagram should include a labelled cell membrane overlying the cytoplasm **[1 mark]**, a thin cell wall overlying this cell membrane **[1 mark]**, and another labelled membrane overlying the cell wall **[1 mark]**. As shown:*

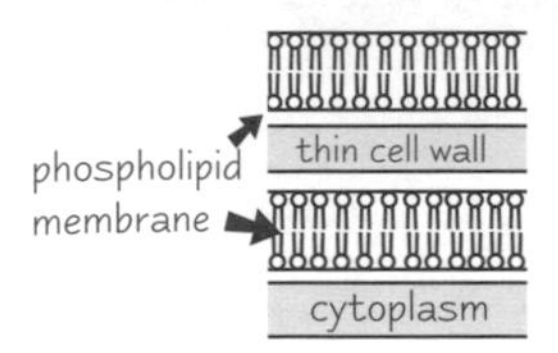

Page 29 — Fungi and Viruses

1 *Maximum of 2 marks available.*
*The latency period is the time between the initial invasion of a cell and the replication of the virus **[1 mark]** when the virus is dormant **[1 mark]**.*

2 *Maximum of 3 marks available.*
*Capsids and envelopes may protect the virus from chemicals whilst outside a host **[1 mark]**, let the virus bind to host cell membranes **[1 mark]** or assist in penetration of a host cell **[1 mark]**.*

Page 31— Cell Culture

1 a) *Maximum of 5 marks available.*
*Hold a wire loop in a flame until red hot and then cool **[1 mark]**. Ensure that there is minimal lifting of the lids from the monoculture and sterile plate **[1 mark]**. Dip the loop into the monoculture **[1 mark]**, then streak the surface of the medium*
***[1 mark]**. One mark for any other precaution, e.g. work close to bunsen flame, wear protective clothing, swab bench with disinfectant **[1 mark]**.*

b) *Maximum of 2 marks available.*
*Heat in an autoclave **[1 mark]** to 121°C for 15 minutes **[1 mark]**.*
Also accept:
The agar plates would be irradiated using gamma radiation.

2 a) *Maximum of 3 marks available.*
*The bacteria respired aerobically **[1 mark]**, which released more energy / made more ATP **[1 mark]** for cell growth / division*
***[1 mark]**.*

b) *Maximum of 2 marks available.*
*There would be no population growth / cells would die **[1 mark]** because oxygen is poisonous to them **[1 mark]**.*

Page 33 — Growth of Cultures

1 a) *Maximum of 3 marks available.*
*$8.6 - 3.4 = 5.2$ **[1 mark]**, $10 \times 0.301 = 3.01$ **[1 mark]**, $5.2 / 3.01 = 1.73$ (generations per hour) **[1 mark]**.*

b) *Maximum of 8 marks available.*
*Take a known / fixed volume of culture at the start of 10 hours / at time = 0 **[1 mark]**. Dilute it by adding water of a known volume / dilute by a known amount **[1 mark]**. Repeat the dilution a fixed number of times **[1 mark]**. Spread a known volume of the final dilution on an agar plate **[1 mark]**. Incubate it **[1 mark]**, then count the number of colonies formed **[1 mark]**. Multiply this number by each dilution factor **[1 mark]**. Repeat for the sample taken at the end of 10 hours / at time = 1 **[1 mark]**.*

2 a) *Maximum of 2 marks available from any of the following.*
*Diauxic growth is when microorganisms are grown on a medium with two different carbon sources **[1 mark]**, so that one carbon source is used before the other **[1 mark]**.*
*The graph of their growth shows two separate exponential phases **[1 mark]**.*

Answers

b) Maximum of 3 marks available.
1 mark for labelling the axes correctly. 1 mark for showing two exponential peaks. 1 mark for correctly labelling lag and exponential phases. As shown:

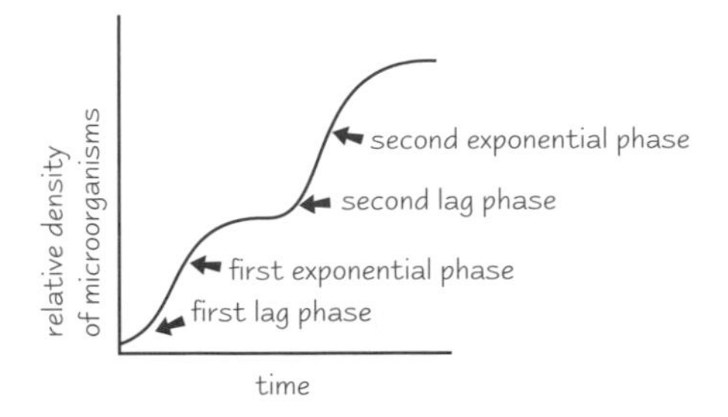

Page 35 — Industrial Growth of Microorganisms

1 a) Maximum of 2 marks available for any of the following:
Batch culture occurs in fixed volume, continuous culture doesn't **[1 mark]**. Medium is added to the continuous culture vessel at a constant rate, but none is added to a batch culture after its start **[1 mark]**. Culture reaches the end of the stationary phase in batch culture, but it's kept at the exponential phase in continuous culture **[1 mark]**.
Batch cultures have to be started over again regularly, unlike continuous cultures which last much longer **[1 mark]**.

b) Maximum of 2 marks available.
Advantages of batch culture: easier to control conditions / less costly to start again in event of contamination **[1 mark]**.
Advantages of continuous culture: method is more productive / culture lasts for longer / a smaller vessel can be used, linked with stated advantage (e.g. easier to sterilise) **[1 mark]**.

c) Maximum of 3 marks available. Possible answers include:
Penicillium would get less oxygen **[1 mark]**, so there would be less aerobic respiration **[1 mark]**, so less ATP would be produced / less growth would occur **[1 mark]**, and less fungus / mycelium produces less penicillin **[1 mark]**.

Page 37 — Biotechnology in Food Production

1 a) Maximum of 3 marks available.
Barley seeds contain starch **[1 mark]** and amylase **[1 mark]**.
Amylase hydrolyses starch into sugar / maltose **[1 mark]**.

b) Maximum of 2 marks available.
Yeast grows / reproduces **[1 mark]** and uses up the oxygen (in respiration) **[1 mark]**.

c) Maximum of 2 marks available.
There would be a higher concentration of sugar / lower concentration of ethanol (alcohol) **[1 mark]** because the rate of fermentation / conversion of sugar to ethanol (alcohol) is less **[1 mark]**.

2 a) Maximum of 4 marks available.
Lactobacillus produces lactic acid **[1 mark]** which gives a sour taste **[1 mark]**. Both produce ethanal **[1 mark]**, which adds to the flavour **[1 mark]**.

b) Maximum of 4 marks available.
Lactobacillus produces peptides (from protein) **[1 mark]** which are used by Streptococcus **[1 mark]**. Streptococcus produces methanoic acid **[1 mark]** which is used by the Lactobacillus **[1 mark]**.

c) Maximum of 2 marks available from any of the following:
Sterilising prevents the addition of other kinds of bacteria **[1 mark]** which could be pathogenic / disease-causing / harmful **[1 mark]** and could spoil the product / flavour **[1 mark]**.

Page 39 — Antibiotics

1 Maximum of 4 marks available. Possible answers include:
A bacterium in the population has a random mutation **[1 mark]** that causes it to produce the enzyme penicillinase **[1 mark]**. This blocks the action of penicillin **[1 mark]**.
The mutant bacteria are more likely to survive when exposed to penicillin **[1 mark]**. The gene for resistance is passed to their offspring and also transferred by conjugation **[1 mark]** via plasmids **[1 mark]**.

2 Maximum of 2 marks available.
Microbicidal antibiotics kill microorganisms **[1 mark]**, while microbistatic ones inhibit (or reduce) their growth **[1 mark]**.

Section 4 — Food Science

Page 41 — Balanced Diet

1 a) Rate of metabolism / energy release when the body is at rest **[1 mark]**.

b) Maximum of 2 marks available.
Metabolism involves respiration **[1 mark]** which uses up oxygen and releases heat energy **[1 mark]**.

c) Maximum of 2 marks available.
More body mass means more respiring tissue / cells **[1 mark]** which releases more energy / uses up more oxygen **[1 mark]**.

2 Maximum of 2 marks available.
Either of the following pairs of points, with 1 mark for each point in a pair: child needs relatively more protein **[1 mark]** to grow **[1 mark]**, child needs more calcium **[1 mark]** for development of bones/teeth **[1 mark]**.

3 Maximum of 8 marks available.
(Dietary) protein is hydrolysed in the gut **[1 mark]** to release amino acids **[1 mark]**. Amino acids are absorbed into the blood **[1 mark]** (taken to cells) and used for protein synthesis in cells **[1 mark]**. Muscle (fibres) are composed of protein **[1 mark]**, so if more protein is synthesised, more can be used for making muscle fibres **[1 mark]**. Excess amino acids are deaminated **[1 mark]** in the liver **[1 mark]** to form urea **[1 mark]**, so more amino acids leads to greater urea production **[1 mark]**.

Page 43 — Under-Nutrition

1 a) Maximum of 4 marks available.
Vitamin C acts as a cofactor / coenzyme (for the enzyme needed to make collagen) **[1 mark]**. The enzyme converts proline into hydroxyproline **[1 mark]** so that hydrogen bonds can form (between hydroxyproline units / residues) **[1 mark]**.
Collagen needs to be replaced for wounds to heal **[1 mark]**.

b) Maximum of 2 marks available from the following:
Citrus fruits **[1 mark]**, green vegetables **[1 mark]**, potatoes **[1 mark]**. Also, allow separate marks for named examples of citrus fruits.

2 Maximum of 4 marks available.
Low protein levels in the blood / plasma **[1 mark]** make water potential higher here / solute concentration lower here **[1 mark]**, so less water reabsorbed back into blood / capillary (from tissue fluid) **[1 mark]** by osmosis **[1 mark]**.

Answers

3 Maximum of 6 marks available.
Less iron in a vegetarian diet **[1 mark]**. Iron in a vegetarian diet is non-haem **[1 mark]**, which is absorbed less effectively than iron from meat / haem iron **[1 mark]**. This means less iron is absorbed into the blood **[1 mark]**, so less haemoglobin is made **[1 mark]**. A shortage of haemoglobin leads to anaemia **[1 mark]**.

Page 45 — Over-Nutrition

1 a) Maximum of 8 marks available.
Cholesterol and saturated fats in the blood (plasma) lead to fatty deposits in artery walls **[1 mark]**, beneath the artery endothelium **[1 mark]**, called atheromas **[1 mark]**. This is calledatherosclerosis **[1 mark]** and it causes narrowing of the arteries **[1 mark]**.
If this happens in coronary arteries **[1 mark]**, it restricts blood flow / oxygen carriage to cardiac / heart muscle **[1 mark]**, resulting in heart attack / myocardial infarction **[1 mark]**.
b) Maximum of 1 mark available for any of the following:
Increased intake of polyunsaturates **[1 mark]**, decreased intake of cholesterol / saturated fats **[1 mark]**, increased intake of soluble fibre **[1 mark]**, increased exercise **[1 mark]**.

2 Maximum of 3 marks available.
Food / faeces stay in the intestine for less time **[1 mark]**, so food / faeces are in contact with the intestine wall for less time **[1 mark]**. This means that harmful substances have less time to affect the wall **[1 mark]**.

Page 47 — Food Additives

1 a) Maximum of 3 marks available.
Sufferers do not produce lactase (in the intestine) **[1 mark]**, so cannot hydrolyse lactose **[1 mark]**. Lactose gets fermented by bacteria in the intestine / gut, causing indigestion **[1 mark]**.
b) Maximum of 2 marks available.
Lactase is added to the food **[1 mark]** to hydrolyse the lactose **[1 mark]** and make the food digestible.

2 a) Maximum of 3 marks available.
Fructose is the sweetest **[1 mark]**, so less fructose is needed to make food as sweet as sucrose would **[1 mark]**. This means fewer calories / less energy is taken in **[1 mark]**.
b) Maximum of 4 marks available.
Immobilised enzymes can be reused **[1 mark]** and are more stable **[1 mark]**. The rate of reaction is higher at high temperatures **[1 mark]** and higher temperatures also reduce growth of microorganisms **[1 mark]**.

Page 49 — Other Additives

1 a) Maximum of 4 marks available.
Flavouring can be tasted **[1 mark]**, examples: spices / herbs / named example, e.g. vanilla **[1 mark]**. Flavour enhancers increase the flavour of other substances **[1 mark]**, examples: monosodium glutamate / sodium chloride **[1 mark]**.
b) Maximum of 6 marks available.
i) High concentration of salt causes microorganisms / cells to lose water by osmosis **[1 mark]** so that they can't grow / multiply / metabolise properly **[1 mark]**.
ii) Vinegar has a low pH / is acidic **[1 mark]** which denatures enzymes / proteins in microorganisms so that they can't metabolise properly **[1 mark]**.
iii) Sulphites lower the oxygen concentration **[1 mark]** so that microorganisms are unable to respire aerobically **[1 mark]**.

Page 51 — Food Storage

1 a) Maximum of 3 marks available.
Pectin makes up middle lamellae (between cells) **[1 mark]**, so pectinase breaks down the middle lamellae **[1 mark]**. This means cells lose their cell sap / juice **[1 mark]**.
b) Maximum of 4 marks available.
Respiration uses up sugar **[1 mark]** so reduces sweetness **[1 mark]**. Ripening converts malic acid to sugar **[1 mark]**, so increases sweetness **[1 mark]**.

2 Maximum of 8 marks available.
Pasteurisation involves heating (milk) to 65°C **[1 mark]**, sterilisation to at least 100 °C **[1 mark]**. Pasteurisation kills off harmful bacteria / does not kill off all bacteria **[1 mark]**, sterilisation kills off more bacteria, but not spores **[1 mark]**. Pasteurised food must be refrigerated **[1 mark]** to reduce the growth / multiplication of bacteria **[1 mark]**, sterilised food does not need to be refrigerated until it is opened, because there will be no viable bacteria **[1 mark]**. Shelf life of sterilised food is longer than that of pasteurised foods **[1 mark]**. Taste of sterilised food may be affected (more than that of pasteurised food) **[1 mark]**.

Page 53 — Microorganisms in Food

1 a) Maximum of 2 marks available.
Salt makes the sauerkraut lose water by osmosis **[1 mark]**. Anaerobic conditions are needed for the bacteria that produce lactic acid and other flavours **[1 mark]**.
b) Maximum of 2 marks available.
In sauerkraut production lactic acid is produced, in wine production ethanol is produced **[1 mark]**. In sauerkraut production bacteria carry out the fermentation, in wine production yeast (a fungus) carries out the fermentation **[1 mark]**.

2 Maximum of 6 marks available.
Fermentation is done by microorganisms acting on starch / flour and boiled soy bean mixture **[1 mark]**. The mould Aspergillus oryzae **[1 mark]** produces amylases / proteases / enzymes **[1 mark]** to release sugars / amino acids **[1 mark: product given must be appropriate for stated enzyme]**. Bacillus / Lactobacillus **[1 mark]** produce(s) lactic acid **[1 mark]**. Yeast / Saccharomyces rouxii **[1 mark]** makes alcohol **[1 mark]**.

Section 5 — Human Health and Fitness

Page 55 — The Cardiovascular System

1 Maximum of 6 marks available.
Cardiac muscle fibres are branched **[1 mark]**. Fibres are connected by intercalated discs in cardiac muscle **[1 mark]**. Striations occur **[1 mark]** caused by overlap of protein filaments **[1 mark]**. Many mitochondria are present **[1 mark]**. Many blood capillaries are present between the muscle fibres **[1 mark]**.

Answers

2 a) *Maximum of 6 marks available from the following:*
Baroreceptors in the wall of the vena cava are stimulated by the increasing blood pressure ***[1 mark]****. An action potential / nerve impulse is sent along the sensory nerve/neurone* ***[1 mark]*** *to the cardioaccelerator centre* ***[1 mark]*** *in the medulla oblongata* ***[1 mark]****. An action potential / nerve impulse is sent along the motor nerve / neurone* ***[1 mark]*** *which is part of the sympathetic nervous system* ***[1 mark]*** *to the SAN, where noradrenaline is released which then increases the heart beat rate* ***[1 mark]****.*

b) *Maximum of 3 marks available.*
Heart would beat irregularly / wouldn't be able to speed up or slow down ***[1 mark]*** *because action potentials / nerve impulses from medulla / cardioaccelerator centre in the medulla would not reach SAN* ***[1 mark]****. In a normally functioning heart these impulses control heart beat rate* ***[1 mark]****.*

Page 57 — The Pulmonary System

1 *Maximum of 4 marks available.*
An action potential / nerve impulse is sent from stretch receptors along the sensory neurone / nerve when the lungs inflate ***[1 mark]*** *to the inspiratory centre of medulla* ***[1 mark]****. The inspiratory centre is inhibited* ***[1 mark]*** *so inspiration stops / expiration begins* ***[1 mark]****.*

2 *Maximum of 3 marks available.*
Thin / squamous epithelium / wall of the alveoli minimises the distance that substances have to diffuse ***[1 mark]****. The large number of alveoli increases the surface area of the lung for diffusion* ***[1 mark]****. The presence of many blood capillaries maximises the concentration gradient of respiratory gasses* ***[1 mark]****.*

Page 59 — The Musculo-skeletal System

1 *Maximum of 10 marks available, for any 10 of the following:*
The sarcoplasmic reticulum membranes become much more permeable ***[1 mark]*** *and calcium ions diffuse out* ***[1 mark]****. They reach the actin filaments and bind to a protein called troponin* ***[1 mark]****, which causes another protein called tropomyosin to change position* ***[1 mark]*** *and unblock the binding sites on the actin filaments* ***[1 mark]****. The myosin heads attach to the binding sites* ***[1 mark]*** *to form actomyosin cross bridges between the two filaments* ***[1 mark]****. The myosin head then changes angle* ***[1 mark]****, pulling the actin over the myosin towards the centre of the sarcomere* ***[1 mark]****. The cross bridges then detach and reattach further along the actin filament* ***[1 mark]****. ATP provides the energy for this* ***[1 mark]****.*

2 *Maximum of 5 marks available.*
More sarcoplasmic reticulum enables action potentials to be carried to more myofilaments ***[1 mark]*** *so more calcium can be released into the cytoplasm* ***[1 mark]****. This means that there is more stimulation of the actin-myosin interaction* ***[1 mark]****, so more of the muscle / more myofibrils are stimulated to contract* ***[1 mark]****. This gives more efficient / faster / stronger muscle contraction* ***[1 mark]****.*

Page 61 — The Lymphatic and Immune Systems

1 *Maximum of 3 marks available from the following:*
The lymphatic system has many lymph nodes ***[1 mark]*** *which contain lymphocytes and neutrophils* ***[1 mark]****. The presence of antigens in the lymph triggers an immune response* ***[1 mark]****. Neutrophils are phagocytic* ***[1 mark]****.*

2 *Maximum of 2 marks available.*
Babies are passively immune to many pathogens ***[1 mark]****. They get antibodies from their mother, through the placenta or breast milk* ***[1 mark]****. Antibody injections give passive immunity* ***[1 mark]*** *in situations where there is not enough time to build up active immunity, e.g. snake bite* ***[1 mark]****.*

Page 63 — Exercise and the Cardiovascular System

1 *Maximum of 2 marks available.*
100 x 48 ***[1 mark]*** *– 4800 cm^3 per minute* ***[1 mark]****.*

2 *Maximum of 3 marks available.*
Myoglobin acts as oxygen 'store' ***[1 mark]****. The oxygen in myoglobin is only offloaded when the oxygen concentration in the muscle gets very low* ***[1 mark]*** *such as during vigorous exercise* ***[1 mark]****.*

Page 65 — Exercise and the Pulmonary System

1 a) *(i) Exercise increases ventilation rate* ***[1 mark]****.*
(ii) Exercise increases the minute volume ***[1 mark]****.*

b) *Maximum of 2 marks available from the following:*
The inspiratory reserve volume is the maximum volume of air that can be breathed in over and above the tidal volume ***[1 mark]****, the value decreases as exercise increases* ***[1 mark]****.*

2 *Maximum of 6 marks available.*
More capillaries develop around the alveoli ***[1 mark]*** *which improves blood flow to the lungs* ***[1 mark]****. This increases the concentration difference of oxygen / carbon dioxide / respiratory gases / makes concentration gradients steeper* ***[1 mark]*** *which increases rate of diffusion* ***[1 mark]*** *of oxygen into the blood* ***[1 mark]*** *and carbon dioxide out of the blood* ***[1 mark]****. This means that more oxygen is carried to the muscles / more carbon dioxide removed from muscles* ***[1 mark]****.*

Page 67 — Exercise and the Musculo-skeletal System

1 a) *Maximum of 3 marks available.*
Insufficient oxygen is delivered to the active muscles ***[1 mark]*** *so anaerobic respiration begins to take place* ***[1 mark]*** *and lactate is a product of anaerobic respiration which cannot be removed from the muscles quickly* ***[1 mark]****.*

b) *Maximum of 2 marks available.*
Oxygen debt is the volume / amount of oxygen needed to get rid of the lactic acid ***[1 mark]*** *that has accumulated in active muscles. The lactic acid is removed by oxidation* ***[1 mark]****.*

2 *Maximum of 3 marks available.*
Taking creatine as a dietary supplement would mean that more phosphocreatine could be made ***[1 mark]****.*
The phosphocreatine donates phosphate to ADP to make ATP ***[1 mark]****. This extra ATP would provide more energy for muscle contraction* ***[1 mark]****.*

Answers

3 Maximum of 6 marks available from the following.
Initially ATP from phosphocreatine is used **[1 mark]** — it is broken down to release phosphate for converting ADP to ATP **[1 mark]**. After that ATP from anaerobic respiration is used **[1 mark]** resulting in production of lactic acid **[1 mark]**; then ATP from aerobic respiration **[1 mark]**; from oxidative phosphorylation **[1 mark]**.

Page 69 — Training

1 a) *Maximum of 5 marks available.*
Anaerobic conditioning is needed in preparation for a sprint **[1 mark]**. This anaerobic training enables the muscles to get more ATP from the two anaerobic systems: the phosphocreatine / ATP-PCr system **[1 mark]** and the lactic acid system **[1 mark]**. Anaerobic training involves brief periods of rest **[1 mark]** to prevent lactate from building up too much **[1 mark]**.

b) *Maximum of 6 marks available.*
Glycogen loading is a method of increasing the glycogen available for muscle contraction in preparation for a specific event e.g. a marathon **[1 mark]**. A period of carbohydrate starvation takes place for a few days ahead of the event **[1 mark]**. This is followed by a diet high in carbohydrate **[1 mark]**. The body then overcompensates and large amounts of glycogen are stored **[1 mark]**. This means that there is a large amount of glycogen in the muscles during the event **[1 mark]** which helps to delay fatigue **[1 mark]**.

2 *Maximum of 3 marks available from the following:*
Trained muscle has more capillaries / a better blood supply **[1 mark]**. Trained muscle has more mitochondria / better ATP production **[1 mark]**. Trained muscle has a larger cross-sectional area **[1 mark]**. Trained muscle has thicker fibres **[1 mark]**. Trained muscle contains more myglobin **[1 mark]**. Trained muscle has more respiratory enzymes **[1 mark]**.

Page 71 — Human Disorders

1 *Maximum of 6 marks available.*
The fats are converted into cholesterol **[1 mark]**. Cholesterol is deposited in the artery walls and causes atheroma / atherosclerosis **[1 mark]**. Blood flow through the arteries is reduced **[1 mark]**. This can put a strain on the heart **[1 mark]** and / or can cut off the blood supply to an area of the heart causing a heart attack **[1 mark]**. Reduced blood flow to the brain can cause localised brain damage resulting in a stroke / cerebrovascular accident **[1 mark]**.

2 a) *Maximum of 2 marks available.*
After the menopause, the deficiency of oestrogen increases the risk of osteoporosis-related bone fractures **[1 mark]**, as parathormone, which is normally inhibited by oestrogen, removes calcium **[1 mark]**.

b) *Maximum of 2 marks available from any of the following:*
In osteoarthritis, bone deformities **[1 mark]** rub against ligaments / soft tissues / other bones **[1 mark]**. Receptors / nerves are stimulated **[1 mark]** and send impulses to the brain **[1 mark]**.

Section 6 — Photosynthesis

Page 73 — Photosynthesis

1 *Maximum of 5 marks available, for any 5 of the following points:*
Leaves are broad and flat (have a large surface area) to absorb as much light as possible **[1 mark]**. They're thin / have a short diffusion distance so CO_2 and water can reach inner cells easily **[1 mark]**. Leaves have veins that contain xylem (to bring water from the roots) and phloem (to carry away the sugars made in photosynthesis) **[1 mark]**.
The cuticle on the upper epidermis protects against UV rays in light and resists dehydration **[1 mark]**. Leaves contain cells with a lot of chloroplasts, which are the photosynthetic factories and contain light absorbing pigments **[1 mark]**.
Air spaces between the cells allow easy diffusion of CO_2 and O_2 gases **[1 mark]**. Stomata allow exchange of gases between the leaf and the atmosphere **[1 mark]**.

2 *Maximum of 5 marks available:*
The stroma contains the enzymes needed for the Calvin cycle (the light-independent stage of photosynthesis) **[1 mark]**. The photosynthetic pigments needed for the light-dependent stage of photosynthesis are found on the chloroplast's thylakoid membranes **[1 mark]**. Thylakoid membranes are covered in stalked particles to make ATP **[1 mark]**. Starch grains found in the stroma act as a storage site for any carbohydrate made by photosynthesis and not used straight away **[1 mark]**. The chloroplast has a double membrane / envelope to allow reactants to be kept close to the reaction site **[1 mark]**. The thylakoids are surrounded by the stroma, so that products of the light-dependent reaction can be quickly used in the light-independent reaction **[1 mark]**.

Page 75 — Factors Affecting Photosynthesis

1 *Maximum of 2 marks available.*
Plants need a constant supply of nutrients for growth, but this doesn't directly affect the rate of photosynthesis **[1 mark]**. The mineral magnesium, however, is important in chlorophyll and so it's essential to light absorption **[1 mark]**.

2 *Maximum of 2 marks available.*
CO_2 is a waste product of respiration, which happens in the mitochondria of plant cells **[1 mark]**. The plant will be using the CO_2 it produces itself for photosynthesis, as well as what it takes in from outside, so just measuring CO_2 uptake by a plant won't give an accurate measurement of the rate of photosynthesis **[1 mark]**.

I've not tested you on them here, but for goodness sake make sure you understand the graphs on p 74. Examiners love them — it's pretty likely that limiting factors will come up somewhere in your exams.

Page 77 — Photosynthesis — The Light-Dependent Reaction

1 a) In the thylakoid membranes of the chloroplasts **[1 mark]**.

b) *Maximum of 2 marks available.*
ATP **[1 mark]** and NADPH + H^+ / reduced NADP / NADPH **[1 mark]**.

c) *Maximum of 3 marks available.*
ATP is produced by both cyclic **[1 mark]** and non-cyclic photophosphorylation **[1 mark]**. NADPH is produced by non-cyclic photophosphorylation **[1 mark]**.

Answers

Page 79 — Photosynthesis — The Light-Independent Reaction

1 a) Ribulose bisphosphate (RuBP) **[1 mark]**.
b) NADPH/reduced NADP **[1 mark]**.
c) Ribulose bisphosphate (RuBP) **[1 mark]**.
d) the enzyme ribulose bisphosphate carboxylase **[1 mark]**.
e) Ribulose bisphosphate (RuBP) **[1 mark]**.

2 a) Between points a and b, light was available and both stages of photosynthesis (light-dependent and light-independent) were happening. ATP and NADPH/reduced NADP were being supplied for the Calvin cycle **[1 mark]**.
b) At point b the light faded and the light-dependent stage of photosynthesis stopped, but the light-independent stage / Calvin cycle continued until point c **[1 mark]**.
c) Photosynthesis stopped at c as supplies of ATP and NADPH were exhausted and no more could be produced **[1 mark]**.

Page 81 — Control of Growth in Plants

1 Maximum of 5 marks available.
Gibberellins stimulate the synthesis of amylase in seeds **[1 mark]** which hydrolyses (stored) starch to maltose **[1 mark]**. Glucose is respired **[1 mark]** to release energy / make ATP **[1 mark]** used in growth of embryo (during seed germination) **[1 mark]**.

2 Maximum of 5 marks available.
Sunlight contains more red light than far red light **[1 mark]**, so exposure to sunlight converts PR into PFR **[1 mark]**. The nights are too short for PFR to be converted back into PR **[1 mark]**. More PFR form builds up in the plant during summer **[1 mark]**. PFR stimulates flowering in long day plants **[1 mark]**.

It's really easy to get mixed up with PR and PFR, or just to write down the wrong one, even if you do know the difference.

Section 7— Biodiversity

Page 83— Classification

1 Maximum of 2 marks available.
Traditional classification deals with features that are easy to observe **[1 mark]** whilst phylogeny deals with the genetic relationships between organisms **[1 mark]**.

2 Maximum of 5 marks available.
a) phylum **[1 mark]**, b) class **[1 mark]**, c) family **[1 mark]**,
d) Aptenodytes **[1 mark]**, e) patagonicus **[1 mark]**.

You will be given all the information you need for this type of question – you don't need to remember scientific names of organisms but you will need to know the groups in the taxonomic system.

Page 85 — Investigating Ecosystems

1 Maximum of 3 marks available.
The light meter would take a measurement for one moment in time **[1 mark]** but light intensity changes over time **[1 mark]**. A single reading would not take these variations into account and would therefore not give an accurate representation of the genuine amount of light in the area **[1 mark]**.

2 Maximum of 2 marks available.
A transect would be used to discover a trend across an ecosystem **[1 mark]**. Any suitable example (e.g. distribution of organisms up a rocky shore, distribution of plants with increasing shade) **[1 mark]**.

3 Maximum of 2 marks available.
Population size = $n_1 \times n_2 / n_m$, 80 x 100/10= 800
[2 marks for correct answer or 1 mark for correct working].

Page 87 — Succession

1 Maximum of 8 marks available.
Succession is a process where plant communities gradually develop on bare land **[1 mark]**. Change goes through stages called seral stages **[1 mark]**. The process stops when a climax community is reached **[1 mark]**. Changes are brought about by the interactions of species **[1 mark]**. Climax may be climatic **[1 mark]** resulting from the climate **[1 mark]** or a plagioclimax **[1 mark]** resulting from human activity **[1 mark]**.

2 a) Maximum of 3 marks available from the points below:
Successful features — rapid growth **[1 mark]**, rapid reproduction **[1 mark]**, asexual reproduction **[1 mark]**, efficient seed dispersal **[1 mark]** tolerant of harsh environmental conditions e.g. high salt levels and strong winds **[1 mark]**.
b) Maximum of 2 marks available.
Reasons for disappearance — shaded by larger plants **[1 mark]**, eaten by herbivores **[1 mark]**, unable to compete for water or minerals with newly arrived species **[1 mark]**.

Page 89 — Populations

1 Maximum of 2 marks available.
Intraspecific competition is where the organisms competing are of the same species **[1 mark]**. Interspecific competition is where they are from different species **[1 mark]**.

2 Maximum of 4 marks available.
A density dependent factor has a greater effect as population density increases **[1 mark]**. A relevant example (e.g. food, oxygen, minerals etc) **[1 mark]**. A density independent factor's intensity is unaffected by the density of a population **[1 mark]**. A relevant example (e.g. fire, flood, drought etc.) **[1 mark]**.

Page 91 — Controlling Pests

1 Maximum of 3 marks available from any of the following:
Non-toxic to humans **[1 mark]**. Specific to weed species being targeted **[1 mark]**. Biodegradable / not persistent **[1 mark]**. Easy to wash off / remove from crops before consumption **[1 mark]**.

2 Maximum of 5 marks available from any of the following:
Biological control is more difficult to apply **[1 mark]**. Biological control can be unpredictable — it's hard to predict all of the knock-on effects **[1 mark]**. Biological control is slower to work than chemical pesticides **[1 mark]**. Biological control does not control sudden outbreaks as well as pesticides **[1 mark]**. Biological control does not completely eliminate the pest **[1 mark]**. Switching from chemical pesticides to biological control will be costly / the farmer may have to purchase new supplies and equipment **[1 mark]**.

Answers

Page 93 — Managing Ecosystems

1 *Maximum of 4 marks available for any of the following:* Growing fast-growing trees **[1 mark]** which will replace themselves quickly **[1 mark]**. Coppicing **[1 mark]** which allows timber to be removed repeatedly without killing the tree **[1 mark]**, increases species diversity **[1 mark]** and allows light to reach the floor level **[1 mark]**.

2 a) Mowing **[1 mark]**.
b) Using the grass/silage cut between May and September **[1 mark]**.
c) It increases biodiversity/allows more species to live in the field **[1 mark]**.

Section 8 — Genetics and Evolution

Page 95 — Variation

1 a) Characteristics that show continuous variation are more likely to be affected by the environment **[1 mark]**.
b) Characteristics that show continuous variation usually involve more genes **[1 mark]**.

2 *Maximum of 2 marks available.*
A homozygous individual has two copies of the same allele for a particular gene **[1 mark]** and a heterozygous one has two different alleles for a particular gene **[1 mark]**.

Page 97 — Genes, Environment and Phenotype

1 *Maximum of 2 marks available.*
2^3 or $2 \times 2 \times 2$ **[1 mark]**, so 8 possibilities **[1 mark]**.
In a question like this, always show your working.

2 a) There is discontinuous variation / there are two distinct categories **[1 mark]**.
b) There is continuous variation within each category **[1 mark]**.

3 *Maximum of 2 marks available.*
Continuous variation **[1 mark]**, because there are no distinct categories **[1 mark]**.

4 Both generations have tall and dwarf plants. This suggests a genetic component as dwarf plants [similar to the parental plants] have appeared in the F_2 generation **[1 mark]**.

Page 99 — Mutation and Phenotype

1 *Maximum of 4 marks available.*
Three nucleotides code for a single amino acid **[1 mark]**. If a whole amino acid is lost the protein / phenotype is not too badly affected **[1 mark]**. If one nucleotide is lost a frame shift **[1 mark]** will cause all the amino acids to change and the protein / phenotype will be severely affected **[1 mark]**.

2 a) *Maximum of 2 marks available.*
To reduce the radiographer's exposure to X-rays **[1 mark]**. X-rays are mutagenic / carcinogenic **[1 mark]**.
b) *Maximum of 2 marks available.*
Sunlight contains UV rays **[1 mark]**, which are mutagenic / carcinogenic **[1 mark]**.
c) *Maximum of 2 marks available.*
The X-rays cause mutation in the parent flies' sex cells / gametes **[1 mark]**. This leads to the production of genetic abnormalities in the next generation **[1 mark]**.

Page 101 — Inheritance

1 *Maximum of 3 marks available — 1 mark for every two correct answers from the following.*
I^AI^A, I^AI^O, I^BI^B, I^BI^O, I^AI^B, I^OI^O

2 *Maximum of 3 marks available.*
Carry out a test cross **[1 mark]**, by crossing with a white-flowered plant **[1 mark]**. If some of the offspring are white-flowered then the plant was heterozygous / If all purple-flowered then the plant was homozygous. **[1 mark]**
This is a really common question in exams, so make sure you answered it correctly.

Page 103 — Inheritance

1 *Maximum of 4 marks available.*

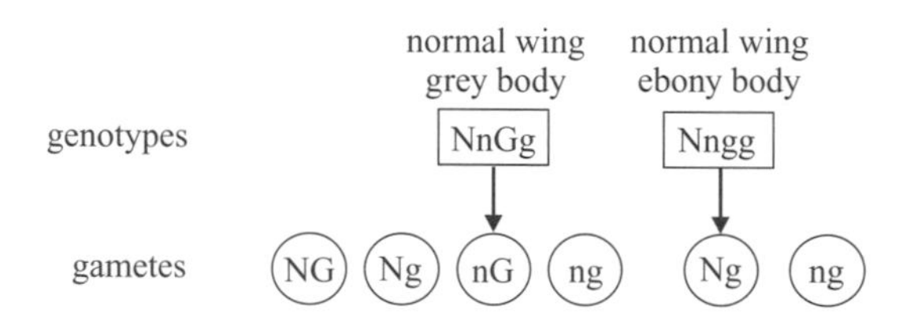

	NG	Ng	nG	ng
Ng	NNGg normal, grey	NNgg normal, ebony	NnGg normal, grey	Nngg normal, ebony
ng	NnGg normal, grey	Nngg normal, ebony	nnGg vestigial, grey	nngg vestigial, ebony

Ratio of 3 normal wings and grey bodies : 3 normal wings and ebony bodies : 1 vestigial wings and grey body : 1 vestigial wings and ebony body.
4 marks for all genotypes and phenotypes correct. Deduct one mark for each mistake.

2 *Maximum of 5 marks available for any of the points below.*
The recessive allele for haemophilia is carried on the X chromosome **[1 mark]**, so males will only have one copy of the allele and females will have two copies **[1 mark]**. This means that a female who inherits one copy of the haemophilia allele will also probably have a copy of the 'normal' allele for factor 8, which is dominant **[1 mark]**. She will be healthy, but a carrier **[1 mark]**. Males who inherit one copy of the haemophilia allele will have the disease **[1 mark]**, so they can get it if they have a healthy father and a carrier mother **[1 mark]**. Females only stand a chance of having haemophilia if they're the child of a haemophiliac male and a carrier female **[1 mark]**, so the probability is much lower **[1 mark]**.

Page 105 — Natural Selection and Evolution

1 *Maximum of 6 marks available from the following points:*
Natural selection has occurred **[1 mark]**. Before the drought individual birds showed variation in beak depth and length **[1 mark]**. When the drought happened birds with longer and deeper beaks had an advantage because they were able to obtain food **[1 mark]**. The birds with the longest and deepest beaks had the greatest chance of survival **[1 mark]**. The birds who survived produced offspring with larger beaks **[1 mark]**. The alleles for larger beaks would therefore be more prominent in future generations of Geospiza fortis on the island **[1 mark]**. This is an example of directional selection **[1 mark]**.

Answers

This question seems complicated because it is really long and it contains a table. Actually, it's not that complicated — it's just asking you to explain how natural selection can change the genetic make-up of a population.

Page 107 — Speciation

1 a) *Maximum of 3 marks available.*
Darwin noticed that there were 14 different species of finch on the Galápagos islands ***[1 mark]****. Each of these finches occupies a different ecological niche* ***[1 mark]****. The finches have different beaks which are adapted to eating different foods* ***[1 mark]****.*

b) *Maximum of 4 marks available.*
Darwin thought that all the finches were originally one species living on one island and competing for resources ***[1 mark]****. Some finches flew to other islands and established separate populations* ***[1 mark]****. Adaptation to different habitat / food sources gradually changed the beak shapes of the finches* ***[1 mark]****. This eventually led to the formation of new species* ***[1 mark]****.*

Section Nine — Genetic Engineering

Page 109 — Genetic Engineering

1 a) *Maximum of 9 marks available.*
The DNA coding for the fluorescent protein is removed from a jellyfish cell ***[1 mark]****. The gene is cut out using restriction endonucleases* ***[1 mark]****. A plasmid is cut open using the same restriction enzymes* ***[1 mark]****. The gene is then spliced into the plasmid* ***[1 mark]*** *using DNA ligase* ***[1 mark]****. The recombinant plasmid is added to a bacterial suspension* ***[1 mark]*** *in ice-cold calcium chloride solution* ***[1 mark]*** *and heated to 42 °C for 50 seconds* ***[1 mark]****, then cooled on ice* ***[1 mark]****.*
Also accept alternative explanation:
Isolate the correct mRNA in a jellyfish cell and use reverse transcriptase to make the complementary DNA ***[1 mark]****. Use the enzyme DNA ploymerase to turn this single-stranded DNA into double stranded DNA* ***[1 mark]****. Insert this DNA into E. coli as described above.*

b) *Maximum of 2 marks available.*
Could be used as a marker gene ***[1 mark]****, as cells containing the recombinant DNA would glow* ***[1 mark]****.*

If you have trouble remembering the sequence of events in transformation, try and make up a clever mnemonic to help.

2 *Maximum of 6 marks available.*
Reverse transcriptase catalyses the production of complementary DNA (cDNA) ***[1 mark]*** *from messenger RNA (mRNA)* ***[1 mark]****. In genetic engineering, it is necessary to find the useful gene to remove, and there are probably only two copies in each cell* ***[1 mark]****, but there will be many mRNA molecules complementary to it* ***[1 mark]****. Finding the mRNA molecules will therefore be easier* ***[1 mark]****.*
The required gene can then be made if nucleotides and reverse transcriptase are put with the mRNA ***[1 mark]****.*

Page 111 — Commercial Use of Genetic Engineering

1 *Maximum of 5 marks available from the following:*
An example of a GM human hormone is insulin / human growth hormone / growth hormone releasing factor (GRF) ***[1 mark]****. The amino acid sequence of the hormone is converted into triplet codons and the gene is synthesised* ***[1 mark]****. A bacterial plasmid is cut with restriction endonuclease* ***[1 mark]****. The gene is spliced into the plasmid using the same restriction enzyme* ***[1 mark]****.*
The recombinant plasmid is introduced into the bacterium ***[1 mark]****. The bacterium is cultured in a fermenter* ***[1 mark]*** *and the product is extracted and purified* ***[1 mark]****.*

2 *Maximum of 6 marks available.*
Benefits *of herbicide-resistant crops —* ***3*** *of: Weeds killed, but not crops* ***[1 mark]****. More potent herbicide can be used, so fewer applications necessary* ***[1 mark]****. Lower cost as fewer applications and less use of machinery, fuel and manpower* ***[1 mark]****. Less compaction of soil and damage of crops as less use of machinery* ***[1 mark]****. Improved yields due to the increased ability to kill weeds* ***[1 mark]****.*
Drawbacks *—* ***3*** *of: Recombinant resistance gene could spread to weed species* ***[1 mark]****. Gene could produce unforeseen side-effects in crops* ***[1 mark]****. Powerful herbicide doses could reduce biodiversity* ***[1 mark]****. Less incentive to research into selective weed killers* ***[1 mark]****.*

You need to do a bit of lateral thinking in questions like this – they're a bit ecological. Don't just put the obvious answers. Remember to give reasons — think A-Level, not GCSE.

Page 113 — The PCR and Genetic Fingerprinting

1 a) *Genetic / DNA fingerprinting / profiling* ***[1 mark]****.*

b) *Maximum of 5 marks available.*
A DNA sample would be taken from the suspect ***[1 mark]****. The DNA from both samples would be cut into fragments by the same specific restriction endonucleases* ***[1 mark]****. The DNA fragments from both samples would be separated out by size using electrophoresis* ***[1 mark]****. Radioactive gene probes would be used to make the DNA fragment patterns of both samples visible* ***[1 mark]****. The bands on the photographic film would be compared to see if the pattern produced by the suspect's DNA matched that produced by the DNA found at the scene of the crime* ***[1 mark]****.*

'Explain' in an exam question tests your ability to apply your knowledge. You must give your answer in the context of the question. Simply describing the technique of genetic fingerprinting wouldn't get you full marks.

Index

Index

Index